Lecture Notes in Computer Science   **8800**

Commenced Publication in 1973
Founding and Former Series Editors:
Gerhard Goos, Juris Hartmanis, and Jan van Leeuwen

Editorial Board

More information about this series at http://www.springer.com/series/8637

Abdelkader Hameurlain · Josef Küng
Roland Wagner (Eds.)

Transactions on Large-Scale Data- and Knowledge-Centered Systems XIV

Springer

Editors
Abdelkader Hameurlain
IRIT, Paul Sabatier University
Toulouse
France

Roland Wagner
FAW, University of Linz
Linz
Austria

Josef Küng
FAW, University of Linz
Linz
Austria

ISSN 0302-9743
Lecture Notes in Computer Science
ISBN 978-3-662-45713-9
DOI 10.1007/978-3-662-45714-6

ISSN 1611-3349   (electronic)

ISBN 978-3-662-45714-6   (eBook)

Library of Congress Control Number: 2009932361

Printed on acid-free paper

Springer-Verlag GmbH Berlin Heidelberg is part of Springer Science+Business Media
(www.springer.com)

Preface

This volume is the fifth so-called regular volume of the TLDKS journal. It contains, 4 fully revised selected regular papers from 12 submitted papers in response to the call for papers for this regular volume.

The content of this volume covers a wide range of different and hot topics in the field of data and knowledge management systems, mainly: data stream systems, top-k query processing, semantic web service (SWS) discovery, and XML functional dependencies.

We would like to express thanks to the editorial board for thoroughly refereeing the submitted papers and ensuring the high quality of this volume.

Special thanks go to Gabriela Wagner for her availability and her valuable work in the realization of this TLDKS volume.

October 2014

Abdelkader Hameurlain
Josef Küng
Roland Wagner

Contents

Reliable Aggregation over Prioritized Data Streams

Karen Works[1]($\boxtimes$) and Elke A. Rundensteiner[2]

[1] Westfield State University, Westfield, MA, U.S.A.
kworks@westfield.ma.edu
[2] Worcester Polytechnic Institute, Worcester, MA, U.S.A.
rundenst@cs.wpi.edu

Abstract. Under limited resources, targeted prioritized data stream systems (*TP*) adjust the processing order of tuples to produce the most significant results first. In TP, an aggregation operator may not receive all tuples within an aggregation group. Typically, the aggregation operator is unaware of how many and which tuples are missing. As a consequence, computed averages over these streams could be skewed, invalid, and worse yet totally misleading. Such inaccurate results are unacceptable for many applications. *TP-Ag* is a novel aggregate operator for TP that produces reliable average calculations for normally distributed data under adverse conditions. It determines at run-time which results to produce and which subgroups in the aggregate population are used to generate each result. A carefully designed application of Cochran's sample size methodology is used to measure the reliability of results. Each result is annotated with which subgroups were used in its production. Our experimental findings substantiate that TP-Ag increases the reliability of average calculations compared to the state-of-the-art approaches for TP systems (up to 91% more accurate results).

Keywords: Data Streaming · Aggregation · Prioritized Resource Allocation

1 Introduction

1.1 Targeted Prioritized Data Stream Systems (*TP*)

Data stream systems process streams of tuples to answer continuous queries. When CPU resources are limited, targeted prioritized data stream systems (*TP*) cannot always process all incoming tuples [6] as motivated below by several application examples. Yet in spite of such overloads, many DSMS must ensure the production of results from certain critical objects. To address these contradicting requirements, TPs utilize application-specific preference criteria to determine

This work is supported by GAANN and NSF grants: IIS-1018443 & 0917017 & 0414567 & 0551584 (equipment grant).
This work started during Karen's Ph.D. study at WPI.

© Springer-Verlag Berlin Heidelberg 2014
A. Hameurlain et al. (Eds.): TLDKS XIV, LNCS 8800, pp. 1–25, 2014.
DOI: 10.1007/978-3-662-45714-6_1

which tuples should be allocated resources ahead of other tuples throughout the query pipeline [2,9,26,36,42–44].

The state-of-the-art TP, Proactive Promotion [43,44], processes more significant tuples ahead of less significant ones throughout the query pipeline. It is a tuple level scheduling approach. Other TP systems use workload reduction approaches, i.e., shedding [2,9,36] and spilling [26,42]. These methodologies remove less significant tuples at the incoming streams. The tuples not removed are processed in FIFO order.

As shown below, the selection of which tuples are processed in TPs is contrary to the production of reliable average calculations. However, in some systems there can be a need to both ensure the production of results from certain critical objects and to generate reliable average results.

1.2   Motivating Examples of TPs

Outpatient Health Care: TP systems track people with dementia [24]. It is critical to monitor people located at improper locations (i.e., likely lost). While monitoring people who live on their own (i.e., need help) may be reduced based on whether or not resources remain after processing people likely to be lost. Until enough resources are available, monitoring any other people could be temporarily skipped. These systems are known to experience data overloads [19].
Military: TP systems track missiles [21]. It may be critical to ensure that each and every object of a certain class is guaranteed to be monitored (e.g., nuclear missiles). While the monitoring of other objects (e.g., missiles bound for unpopulated areas) may be reduced based on whether or not processing resources remain after processing more significant objects. In addition monitoring of certain objects could in the worst case be temporarily skipped altogether (e.g., missiles sent by our country) until all other objects can be processed within their response time.
Law Enforcement: TP systems monitor prisoners assigned to home arrest [13]. They also get overloaded. In October 2010, an application monitoring released sex offenders across 49 states shut down for 12 hours [32]. With the highest level of urgency, violent prisoners (i.e., may cause harm) must be monitored. Next, prisoners at an improper location (i.e., likely to be in violation) shall be monitored. Finally, if resources permit, prisoners known to be flight risks ought to be monitored.

1.3   Running TP Example: Stock Market

TP systems monitor stocks online [20]. Such applications can get overloaded. In 2012, the London Stock Exchange shut down after a rash of computer-generated orders overwhelmed the system [30].

Consider a data stream application that monitors the average price of stocks by their business sector that appear in recent news and blogs (*Q1* below). Results are formed when news tuples join with stock tuples based upon their business sector (op1). Then these join results are joined with blog tuples based upon their business sector (op2). Finally, the average price for every business sector of these join results are created (op3).

(Stock Market Query)	/*Operators*/
Q1:SELECT AVG(S.price)	/*op_3*/
FROM Stock as S, News as N, StreetResearch as SR	
WHERE contains(S.sector, News[10 min])	/*op_1*/
AND contains(S.sector, StreetResearch[15 min])	/*op_2*/
Group by S.sector	
WINDOW 30 sec;	

Mutual fund companies often invest in diverse stock portfolios. It is critical to ensure that every tuple of a certain class (e.g., their aggressive investments) is processed. While the processing of other tuples (e.g., their conservative investments) may be reduced based on what resources remain after processing the more significant tuples. Until there are enough resources to process all important tuples, monitoring of certain tuples can be temporarily skipped (e.g., stocks under evaluation). When the Stock Market Application is extremely overloaded, the scarce CPU resources will be dedicated only to the tuples most critical for the application, namely, tuples from aggressive investments.

1.4 Running Example: Inaccurate Aggregation Results

Consider the Stock Market Aggregation Example in Figure 1. Tuples from the stock, news and street research streams are respectively depicted by circles, squares, and triangles. The significance of each tuple is represented by its color. Black, grey, and white are respectively the most (i.e., level 1), the average (i.e., level 2), and least (i.e., NA) significance levels. In Figure 1 the system is overloaded. No tuples at level NA arrive at the aggregate operator op_3. The state of aggregate groups g_1 and g_4 thus only contain data from tuples at levels 1 and 2.

This may cause the average price per business sector produced by aggregate operator op_3 in query Q1 to be skewed. Clearly, under limited resources some tuples may not be used to create the aggregate result. Rather in this case, the aggregate result will be generated only from those most significant tuples that reach operator op_3. Unfortunately, this result may not match the aggregate result that would have been produced if all tuples had been processed by the aggregate operator. This makes the result unreliable and worst yet misleading. In our particular example, a broker would thus not know which subset of tuples was used to compute each aggregate value. Without such knowledge, a broker may make poor trades, which may lead to investors losing money.

At the very least we would want to inform application users about which tuples were used to create the result. Consider the example above. The aggregate result should be annotated with the fact that it is formed from aggressive investments only (i.e., a particular subgroup of the overall population) instead of from all investments. However, sometimes there may not even be enough

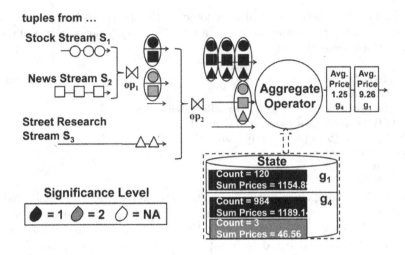

Fig. 1. Stock Market Aggregation Example

resources to process all tuples from a subgroup. That is, the subgroup may be incomplete. Aggregate results produced from such subgroups may again risk to be misleading and faulty. Thus, an aggregate operator must be designed to only generate results from subgroups that reliably represents their actual population.

As shown below, state-of-the-art aggregation methods for TPs [4,39] do not tackle this problem. This critical problem of reliable average calculation from incomplete populations is now the focus of our work.

1.5 State-of-the-Art Aggregation Operators and Their Shortcomings

The state-of-the-art aggregation methodologies [4,39] designed to produce reliable results in TPs address the problem by trying to solve the issue of system overload. They adapt the selection of which tuples are processed and which are not. [39] only processes tuples from windows where the population is guaranteed to be complete. Tuples from other windows are not processed. [4] adapts how many tuples are randomly dropped to ensure that all aggregate results produced are bound by a given error rate. These approaches adapt which tuples are allocated resources. By changing how resources are allocated, they may not abide by the user defined resource allocation preferences. This adaption assumes that accuracy is more important than the user-defined resource allocation preferences which may not always be the case.

Olston et al. [31] proposed a system that creates an aggregate result from cached results and the actual data set with the goal of creating an aggregate result within a high confidence interval range as quickly as possible. They propose a methodology to determine which results to cache that considers the trade off between precision and performance. Their algorithm delivers an answer that is bound by a specified precision constraint. They do not address that the data may be skewed by the set of tuples that are selected to be processed by the TP. Their approach assumes that all tuples within a population are given equal chance at reaching the aggregate operator.

It is challenging to construct reliable aggregate results from only the tuples that arrive at the operator. Consider the stream aggregate operator op_3 [10] in a TP under duress that implements the Stock Market query Q1 above (Fig. 1). The average price for business sector group g_4 could be produced from the tuples at levels 1 and 2. There are only 3 tuples at level 2. Such few tuples may not be representative of the actual population of tuples at level 2. TP requires an aggregate operator than can selectively control which subgroups are used to generate each result. This raises the question of how to determine which subgroups best represent their actual populations.

1.6 Our Approach and Contributions

We propose the TP aggregate operator *TP-Ag* to address the issues described above. TP-Ag produces reliable aggregate average results from incomplete subgroups. It uses an efficient estimation model to determine how representative each subgroup is of their actual population. In addition, TP-Ag annotates each aggregate result with the subgroup population(s) that the result is generated from.

The design of an aggregate operator that meets the requirements above is challenging. TPs pulls specific significant tuples ahead of other tuples. Some TPs only identify significant tuples upon arrival at the data stream system [2,9,26,36,42], while others can identify significant tuples at operators further down the query pipeline [43,44]. This makes it more complicated to determine which tuples from specific populations never reach the aggregate operator. It is equally challenging to estimate what contribution these tuples would have had on any aggregate result produced.

Our Contributions Include

- We formulate the TP aggregate operator problem of generating reliable annotated aggregate results from incomplete populations.
- We propose a carefully designed estimation model and application of Cochran's sample size methodology to measure if any subset of the actual population is large enough to generate a reliable aggregate result. This requires knowledge of how many tuples fail to reach the aggregate operator due to limited resources. We propose a method to track this information.
- We carefully outline the logical design of our novel TP-Ag operator, namely, how TP-Ag generates reliable aggregate results by selectively controlling which subsets in the aggregate group population are used to generate each result.

- We carefully outline the physical design of our novel TP-Ag operator. The data structures and infrastructure behind our TP-Ag operator are designed to allow efficient retrieval of subgroups within an aggregate population.
- Our experiments show that TP-Ag produces up to 90% more accurate aggregate results compared to the state-of-the-art methodologies on a wide variety of data sets and workloads.

2   TR Query Model and Plan Definitions

2.1   TP Queries

A query q_j for a TP is a CQL query [3] extended to allow the specification of multi-tiered monitoring criteria to identify significant tuples in TP systems. Below is the *Stock Market Query Q1* with the extension to support TPs.

(Stock Market Query with Extension)   /*Operators*/
Q1:SELECT AVG(S.price)   /*op_3*/
   FROM Stock as S, News as N, StreetResearch as SR
   WHERE contains(S.sector, News[10 min])   /*op_1*/
   AND contains(S.sector, StreetResearch[15 min])   /*op_2*/
   Group by S.sector
   WINDOW 30 sec;   **RANK** 1 /* aggressive investments */
   CRITERIA (S.ownedByCompany=TRUE) **AND** (S.aggressive=TRUE)
   RANK 2 /* conservative investments */
   CRITERIA (S.ownedByCompany=TRUE) **AND** (S.conservative=TRUE)
   RANK 3 /* stocks under evaluation */
   CRITERIA (S.underEvaluation= TRUE)

Monitoring levels $q_j.ML$ are predicates defined at compile-time that state the user's desired result production order. When resources are scarce, the monitoring levels are used to identify which tuples will be processed (i.e., not shed). Each monitoring level m_k states m_k's *rank* $m_k.rnk$ and *membership criteria* $m_k.mem$. The rank $m_k.rnk$ denotes how significant monitoring level m_k is compared to the other monitoring levels. When resources are limited, TP optimizers [2,9,26,36,42–44] identify which types of tuples to process based upon the monitoring levels. These tuple are called *significant tuples*.

The *query window* denotes the finite set of tuples from stream s_n used to create results. *Window bounds* are specified as time (e.g., within a 30 second time span) or count (e.g., last 30 tuples) ranges. In *sliding windows*, tuples from consecutive windows may overlap. While we assume time-based sliding windows, our proposed techniques equally apply to count-based window semantics.

2.2   TP Query Plans

A TP query plan is a directed acyclic graph composed of *TP query operators* as nodes and *data exchange interfaces* that transfer tuples between operators as edges. The *data exchange interface* transfers tuples between operators.

TP operators include enhanced traditional operators and significance classifiers. Traditional operators from the continuous query algebra [14] are enhanced to allocate resources to incoming tuples in significance order and to propagate significance properties assigned to tuples. Significance classifiers (or *SCs*) are special-purpose operators that compute and assign significance properties to tuples [43,44].

3   Background of Stream Aggregation

3.1   Basic Aggregate Operator

The *Basic Aggregate Operator* [8] computes a function over the set of tuples that belong to the same aggregate group within the current query window w_p of the stream. In Stock Market Query Q1, the aggregate group is a business sector. Incoming tuples are stored in the state and associated with their aggregate group. When it is determined that no future incoming tuples for window w_p will arrive then the aggregate result(s) are generated for each aggregate group in w_p.

3.2   Aggregate Operator Supporting Out-of-Order Data Streams

Some TPs [43,44] cause tuples to become out-of-order. TP aggregate operators require special support to know when all tuples from a window have been processed to produce their results. An aggregate operator should produce results for window w_p when no tuples from window w_p remain to be processed. Prior out-of-order work [23] proposed to use punctuations to trigger the creation of aggregate results whenever a window is complete. To quickly identify tuples within a given window, windows are divided into groups of tuples (a.k.a. *panes*) [22]. Each pane p_q is assigned a pane number. When a tuple arrives, it is associated with a pane number which determines the windows the tuple will produce results for. Each aggregate result is only generated from tuples whose panes compose its query window. Each operator contains a punctuation queue that stores notifications described below.

To discern when no tuples from pane p_q remain to be processed, the progress of tuples is tracked. Each leaf operator op_o periodically sends a punctuation into when it has processed all tuples from pane p_q. To send a punctuation requires simply placing the punctuation into the punctuation queue of the next down stream operator. Upon receiving such a notification, the aggregate operator produces all results for the window that ends with pane p_q. The Out-of-Order Aggregate Operator assumes that tuples are processed. It can produce unreliable results.

3.3   State-Of-the-Art Aggregate Operator for TPs

The *State-Of-the-Art TP Aggregate Operator* (Sec. 1.5) also produces unreliable results. The operator contains a punctuation queue that stores notifications (See above) and incoming queues that store tuples by significance level (Sec. 2). It works as follows.

Producing Aggregate results: Aggregate results are created for the first notification in the punctuation queue and are sent to the next operator (line 4). Finally, tuples stored in the state within panes that will not produce any future results are purged (line 5). This continues until either no resources or incoming punctuations remain (line 1).

Processing tuples: If resources remain then starting from the most significant incoming queue tuple t_i is stored and associated with the proper aggregate group and pane (line 11). After all tuples from one queue have been processed, tuples in the next significant incoming queue are processed (lines 13-16). This continues until either no resources remain or all queues are empty (line 8).

```
Algorithm State-Of-the-Art TP Aggregate Operator(
        Qp         /pane complete punct. queue/,
        Cavail     /avail. res./,
        Qinc(s1) /queues for stream s1/)
        Cavail     /avail. res./,
        Qinc(s1) /queues for stream s1/)
1:  WHILE ((Cavail > 0) and (Qp is not empty))
2:     Punc = first pane complete punctuation in Qp
3:     create aggregate results that for window that ends with pane in Punc
4:     place aggregate results into input queue of next query plan operator
5:     purge state of tuples within panes that will not produce any future results
6:  ENDWHILE
7:  LevelProcessed= 1
8:  WHILE((Cavail > 0) and (Qinc(s_1) is not empty))
9:     IF(Qinc(s_1,LevelProcessed) contains a tuple)
10:     Tup = first tuple in Qinc(s_1,LevelProcessed)
11:     store Tup in state and associate with the aggregate group and pane
12:   ELSE
13:      LevelProcessed = LevelProcessed + 1
14:      IF (LevelProcessed > max(Monitoring Level))
15:         LevelProcessed= Insignificant Tuple
16:      ENDIF
17:   ENDIF
18: ENDWHILE
```

4   TP-Ag Problem Definition

Our TP-Ag operator must meet the following requirements.

1. It must produce the most reliable aggregate result from the tuples that arrived at the aggregate operator and belong to selective subgroup populations. The set of tuples used to create each result must be estimated to represent the actual selected subgroup populations of tuples that would have arrived at the aggregate operator if resources were available.

2. It must annotate each aggregate result with the subgroup population(s) that the result is generated from.

3. It cannot ignore the desired resource allocation order specified by the user and adjust which tuples the TP optimizer chose to process (Sec. 2). It must build reliable aggregate results from the significant tuples already pulled forward. Data streams are often shared to efficiently produce more than one output. Adjusting the desired resource allocation order specified by the user assumes that the aggregate result is the most important query result produced (which may not be the case).

5   TP Aggregation Foundation

5.1   Aggregate Result Annotation

Per the requirements (Sec. 4), each aggregate result must be annotated with which tuples it is generated from. In TP, tuples chosen to be processed satisfy the membership criteria of an activated monitoring level (Sec. 2). Thus we propose to logically divide the population of tuples within an aggregate group and pane into subsets based upon their significance levels. Each population is divided into subsets, namely, one subset for each significance level and one for insignificant tuples.

Each aggregate result a_i is associated with a population generation subset flag or ss_l that annotates which tuples the aggregate result a_i is generated from. ss_l is represented as a bit vector with one bit for each monitoring level $lvl_1, lvl_2, ..., lvl_n$ and one bit for insignificant tuples. For instance, subsets flag $ss_1 = 100$ signifies that the aggregate result a_i is generated from only tuples at significance level 1. While subsets flag $ss_2 = 110$ signifies that the aggregate result a_i is generated from only tuples at significance levels 1 and 2. Finally, subsets flag $ss_3 = 111$ signifies that the aggregate result a_i is generated from all tuples (both significant and insignificant).

Traditionally, aggregate operators produce a single aggregate result for each group and window. It is possible for each subset of significant tuples in an aggregate group and window to create an aggregate result. Given the limited resources, we propose to follow the former method. The goal of TP-Ag is to produce the single most reliable aggregate result from the largest number of subsets for an aggregate group. To achieve this, TP-Ag selectively choses which of the available subset(s) are used to create each aggregate result.

5.2   Evaluation Strategies for Sample Population

Result Accuracy: The actual population pop_m is the set of tuples in the aggregate group g_l and window w_k that given adequate resources would have reached aggregate operator op_o. Aggregate answer a_i^* is generated from an actual population pop_m. Aggregate result a_i is generated from a sample population $spop_m$, i.e., the set of tuples that reached aggregate operator op_o. The sample population is a subset of the actual population. Aggregate result a_i may be incorrect, i.e., not match the aggregate answer a_i^*. Reconsider Figure 1. A sample population $spop_1$ for aggregate group g_4 includes 987 tuples where 984 tuples have significance level 1 and 3 tuples have significance level 2.

Determining Sample Population Accuracy: One method to determine if a sample population accurately portrays the actual population is to compare the mean of the sample and actual population via the Hoeffding equation [17]. Many aggregation operators [16,23,31] that process aggregate results from most (if not all) tuples in a given query window (Sec. 8) use this approach. TP-Ag seeks to build reliable aggregate results solely from the tuples processed by controlling which of the available subgroup(s) are used to create an aggregate result. However, TP-Ag cannot use the Hoeffding equation to determine the accuracy of the sample population. The Hoeffding equation requires that the actual mean μ be precisely measured. TP cannot ensure that all tuples reach TP-Ag op_o. Thus to calculate the actual mean μ requires knowledge of which tuples could have reached TP-Ag op_o but did not. Such an evaluation amounts to running the full query. Clearly, this is prohibitively costly.

Thus lead us to look at other statistics methods. One method to determine if a sample population accurately portrays the actual population is to estimate the sample size required to determine the actual mean within a given error threshold via Cochran's sample size formula [7]. If the size of sample population, denoted by $|spop_m|$, is less than the estimated required sample size $|spop^{est}|$, then the sample population $spop_m$ may not accurately represent the actual population pop_m. No aggregate results should be generated from $spop_m$. TP-Ag can use Cochran's sample size formula [7]. It determines the sample size by considering the limits of the errors in the mean values of items in the sample population. $|pop_m|$ is the size of the actual population. ϵ is the user selected error rate. σ is the standard deviation of the actual population. z is the user selected confidence level or the estimated percentage of the values in the sample population within two standard deviations of the mean of the actual population. The Cochran's sample size formula [7] is $|spop^{est}| = (z^2 * \sigma^2 * (|pop_m|/(|pop_m| - 1)))/(\epsilon^2 + ((z^2 * (\sigma^2)/(|pop_m| - 1))))$. Roughly, $z^2 * \sigma^2 * (|pop_m|/(|pop_m| - 1)))$ represents the percentage of tuples from the sample population that are within the confidence interval of the estimated mean. $(\epsilon^2 + ((z^2 * (\sigma^2)/(|pop_m| - 1))))$ represents the percentage of tuples from the sample population that per the error rate must be within the confidence interval of the estimated mean.

This approach requires that the aggregate values in the population follow a normal distribution. While not all data has this distribution, many practical streams do. One example is stock market prices which can be mapped to the normal distribution [11].

We must calculate the standard deviation of the actual population σ and the size of the actual population $|pop_m|$. As common practice, we propose to calculate the standard deviation of the actual population σ using the standard deviation of the sample population and Bessel's correction [18]. To estimate the actual population size $|pop_m^{est}|$, we must measure *the estimated number of expired tuple* or how many tuples would have reached TP-Ag operator op_i if given adequate resources. To calculate the estimated number of expired tuple the number of tuples that expire at each operator in the query path before aggregate operator op_o (i.e., $|exp(op, g_l, lvl_p)|$) and the probability of expired tuples reaching aggregate operator op_o (i.e., $P(op, op_o, lvl_p)$) is tracked (Sec. 6). The estimated number of expired tuple is the product of these two values (i.e., $P(op, op_o, lvl_p)*|exp(op, g_l, lvl_p)|$).

5.3   Policy for Selecting the Sample Population

There are many ways of selecting which subgroup(s) are used to generate a result. We propose to include the largest number of reliable subgroups in consecutive significance order in the sample population. For example, aggregate result a_i could be created from only tuples at significance level 1, at significance levels 1 or 2, or all levels.

Reconsider the aggregate operator op_3 (Fig. 1). An aggregate result for aggregate group g_4 could be created from the two subgroups, namely, the 984 tuples at significance level 1 and 3 tuples at significance level 2. Assume that the estimated actual population size $|pop_m^{est}|$ for this sample population is 1984 tuples. The estimated standard deviation σ is 7.9. The error rate ϵ is .05. The critical standard score z is 1.96 (95% confidence level). Then the estimated required sample size $|spop^{est}|$ (i.e., $= (1.96^2 * 7.9^2 * (1984/(1984 - 1)))/(.05^2 + ((1.96^2 * (7.9^2)/(1984 - 1))))$) is 1943 tuples. This sample population is not large enough to create a reliable aggregate result.

However, an aggregate result could be created for group g_4 that only using the 984 tuples at significance level 1. Assume that the estimated actual population size $|pop_m^{est}|$ for this sample population is 1000 tuples. The estimated standard deviation σ is 5.9. The error rate ϵ is .05. The critical standard score z is 1.96 (95% confidence level). Then the estimated required sample size $|spop^{est}|$ (i.e., $= (1.96^2 * 5.9^2 * (1000/(1000 - 1)))/(.05^2 + ((1.96^2 * (5.9^2)/(1000 - 1))))$) is 981 tuples. This sample population is large enough to produce a reliable aggregate result.

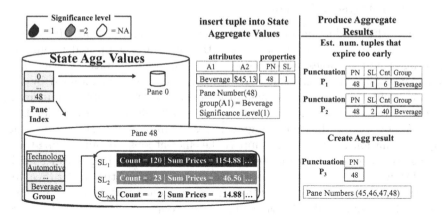

Fig. 2. TP-Ag State Example

6   Design of the TP Aggregation Operator

6.1   Tracking Expired Tuples

We designed the traditional operators (Sec. 2) to periodically send an *expiration count punctuation* to the TP-Ag operator. The expiration count punctuation contains the number of expired tuples for each aggregate group and significance level. TP-Ag uses these values to estimate the actual population size (Sec. 5.2).

Operators track the number of tuples that expire too early over a window by aggregate group and significance level (Sec. 5.2) and send this count to the TR-Ag operator. TR-Ag will use this count to estimate the actual population size when deciding whether or not a sample population should produce a result. To achieve this, TR-Ag uses an *expiration count punctuation*. Each operator tracks the estimated number of tuples that expire too early by pane, aggregate group, and significance level. When no tuples in a pane remain to be processed by an operator, the operator sends an expiration punctuation for each aggregate group and significance level where tuples expired. Upon receiving an expiration punctuation, each operator adds on their estimated number of tuples that expire too early. The expiration punctuation moves along the pipeline until it reaches the TP-Ag (Sec. 6.2). When, the expiration punctuations reach the TP-Ag they contain the number of tuples that expire too early over a window for each aggregate group and significance level.

Consider expiration punctuations $p_1 < 48, 1, 6, Beverage >$ and $p_2 < 48, 2, 40, Beverage >$ in Figure 2. Expiration punctuation p_1 states that 6 tuples from pane 48, significance level 1, and group Beverage are estimated to have expired too early. While expiration punctuation p_2 states that 40 tuples from pane 48, significance level 2, and group Beverage are estimated to have expired too early.

6.2   TP-Ag Physical Design

To support the production of aggregate results from certain subset(s) of the sample populations, TP-Ag must support the efficient look-up and purging of stored aggregate variables by pane (Sec. 3), group, and significance level. The number of tuples within each pane is limited. Thus tuples maintained by the operator are first grouped by the panes they belong to. Next, tuples are indexed by their aggregate group. Lastly, tuples are stored by their significance level.

To support the production of aggregate results from certain subset(s) of the sample populations, TP-Ag must support the efficient look-up and purging (Sec. 6.4) of stored aggregate variables by pane (Sec. 3), group, and significance level. **State Design:** TP does not always process tuples in arrival time order (Sec. 3). Thus, the state can contain tuples from more than one query window. That is, a multitude of tuples with the same aggregate group and significance level are likely to exist across multiple panes. However, the number of tuples within each pane is limited. Thus stored tuples are first grouped by the panes they belong to. Next, tuples are indexed by their aggregate group. Finally, tuples are stored by their significance level.

Consider the insertion of stock tuple $t_1 < Beverages, \$45.13 >$ with pane 48 and significance level 1 into state aggregate values (Fig. 2). Stock tuple t_1 is stored in the state for pane 48, business sector *Beverages*, and significance level 1.

6.3   TP-Ag Operator

The major difference between TP-Ag and the State-Of-the-Art TP Aggregate Operator is that TP-Ag upon receiving a notification punctuation (Sec. 3) determines the largest number of reliable subgroups in consecutive significance order in the sample population that can produce a result (if any). Consider the production of aggregate results triggered by punctuation $p_2 < 48 >$ (Fig. 2). p_2 signals that all tuples from pane 48 that arrived at the operators that reside in the query pipeline prior to the aggregate operator have been processed. First, TP-Ag locates the group attributes for pane 48 which are Technology, Automotive, ..., and Beverage. Then for each group attribute (e.g., Beverage), it creates a sample population from all tuples from the window (composed of 4 panes) that ends with pane 48 (i.e., pane 45 – 48). If the size of the sample population is greater than or equal to the estimated required sample size then TP-Ag creates an aggregate result for tuples from this sample population. Otherwise, TP-Ag creates new sample population by removing the least significant subset from the current sample population. The process of testing the current sample population and creating a new sample population continues until either a result is generated or the sample population is empty. In the latter case, no aggregate result will be produced. Then TP-Ag moves to the next group for pane 48 (and so on...).

It works as follows. First, the incoming punctuations are processed.

Processing Expiration Punctuations: Any incoming expiration punctuation is stored in the state (lines 3-6).

Producing Aggregate Results: Any incoming notification punctuation is processed as follows. First, the groups in the window pane are identified (line 7). Next, for each group, we test to see if the sample population for all tuples represents the actual populations (line 10). If so, results are created and sent to the next operator (lines 11-12). Otherwise, the sample population is reduced by tuples that belong to the least significant subset (line 14) and the test starts over (line 10). This continues until either a result is produced or all sample populations have been explored (line 8). Then, results for the next group are produced. This continues until either no resources remain or there are no more incoming punctuations (line 1). Finally, tuples stored in the state within panes that will not produce any future results are purged (line 18).

Processing tuples: This is the same as the processing tuples logic for State-Of-the-Art TP Aggregate Operator (Sec. 3) (lines 18-29)

```
Algorithm TP-Ag Operator( Qep      /exp. punct. queue/,
                          Qnp      /not. punct. queue/,
                          Cavail   /avail. res./,
                          Qinc(s_1) /queues for stream s1/)

1: WHILE((Cavail > 0) and (Qnp is not empty))
2:   Punc = first punctuation in Qnp
3:   WHILE((Qep is not empty) and (first punctuation in Qep = pane in Punc))
4:     ExpPunc = first punctuation in Qep
5:     store the values in ExpPunc
6:   ENDWHILE
7:   FOR each group gl in defined by the pane in Punc
8:     WHILE(no result has been produced) and (the sample population is not empty)
9:       sample population = population defined by the pane in Punc and group gl
10:      IF (sample population represents actual population)
11:        create aggregate results for sample population
12:        place aggregate results into input queue of next query plan operator
13:      ELSE
```

```
14:      reduce the sample population by the least significant subgroup
15:     ENDIF
16:    ENDWHILE
17:  ENDFOR
18:   purge state of tuples within panes that will not produce any future results
19: ENDWHILE
18: LevelProcessed= 1
19: WHILE((Cavail > 0) and (Qinc(s_1) is not empty))
20:  IF(Qinc(s_1,LevelProcessed) contains a tuple)
21:    Tup = first tuple in Qinc(s_1,LevelProcessed)
22:    store Tup in state and associate with the pane, aggregate group, and significance level
23:  ELSE
24:    LevelProcessed = LevelProcessed + 1
25:    IF (LevelProcessed > max(Monitoring Level))
26:      LevelProcessed = Insignificant tuples
27:    ENDIF
28:  ENDIF
29: ENDWHILE
```

6.4   Memory Resource Management

Beyond CPU resources, memory resources may also be limited.

State Management: To ensure complete results, tuples stored in states are not purged if they may create aggregate results in the future (Sec. 3). However, this purging method assumes that sufficient memory is available to store all tuples that will create future aggregate results. This may not always be the case. In the case of insufficient memory, we propose to purge tuples from the oldest panes in the state first. Our approach is based upon the fact that the majority of aggregate results generated from the oldest tuples would have already been produced. Memory resources are allocated to storing the freshest tuples.

Queue Management: In the case of insufficient memory, the incoming queues must also be purged. We also utilize the oldest pane method defined above to purge the queues.

7   Experimental Evaluation

7.1   Experimental Setup

Alternative Solutions. We compare *TP-Ag* (or *TP w/ TP-Ag*) to the state-of-the-art aggregate operators in TPs. That is, we compare to the out-of-order aggregate operator [23] implemented in the state-of-the-art TP tuple level scheduling approach (or *PP*) [43,44] and the stream aggregation operator [8] implemented the state-of-the-art TP workload reduction approaches, namely, semantic (or *sem*) and random (or *rand*) [2] (Sec. 8). PP requires the out-of-order aggregate operator (Sec. 3). We also compare TP w/ TP-Ag to state-of-the-art aggregate operators for TPs [4,39] that limit which tuples are dropped from specific windows (or *Shed Window Ag*). Finally, we compare to the traditional aggregate operator [8] implemented in a non-targeted prioritized data stream systems (or *trad*). TP-Ag uses the critical standard score $z = 1.96$ (95% confidence level). To ensure fairness, all systems are implemented in the same data stream system with appropriate extensions to implement the methods, in our case, CAPE [34].

TP w/ TP-Ag, PP, and sem use the same criteria to select the tuples processed. Rand randomly selects tuples to process in FIFO order based upon the estimated number of tuples that can be processed within their lifespan. Trad simply processes all tuples in FIFO order.

Data Streams and Query. Most experiments use Stock Market Query Q1 with the extension (Sec. 2) where the window size = 500 tuples.

The stock market stream was created from stock ticker information on the S&P 500 stocks gathered over July 18, 2012 via Yahoo Finance [12].

News and blog data streams were created by randomly selecting sectors from the global industry classification standard (GICS). GICS, developed by Morgan Stanley Capital International (MSCI) and Standard & Poors, contains 10 sectors that categorize the S&P 500 stocks.

Data Set 1 (or *DS1*) mimics a financial company monitoring diversified mutual funds. That is, the stocks chosen are distributed across different business sectors and investment types (i.e., aggressive versus conservative investments). In DS1, 5% of the 500 stocks (or 25 stocks) were randomly selected to be at each of the three monitoring level (Sec. 2.2).

Hardware. All experiments are conducted on nodes in a cluster. Each host has two AMD 2.6GHz Dual Core Opteron CPUs and 1GB memory.

Metrics and Measurements. TP w/ TP-Ag produces an aggregate result from a subset of the tuples that arrive at the aggregate operator. The other approaches generate aggregate results from all tuples that arrive at the aggregate operator. To be able to validate whether or not the results are correct, each aggregate result produced is annotated with the significant levels of the tuples in the sample population. For each experiment, the actual aggregate answers (Sec. 5.2) for each possible sample population was found.

The experiments were run 3 times for 10 minutes. The results are the average of these runs. Each aggregate answer produced is then compared to the actual aggregate answer for the same group, window, and sample population. Any result that is within 5% of the actual answer is considered to be correct. Our experiments measure the percentage of correct aggregate results produced.

7.2   Experimental Methodology

We explore the following: 1) Is TP-Ag more effective at producing a larger percentage of correct significant aggregate results than the state-of-the-art solutions? 2) What effect does the number of significant tuples that belong to each aggregate group have on the effectiveness of the TP-Ag strategy compared to the state-of-the-art solutions? 3) How do changes in the error rate (Sec. 5.2) affect the percentage of correct results produced by TP-Ag? 4) What is TP-Ag's runtime CPU and memory overhead in the worst case scenario compared to the state-of-the-art solutions?

We vary the *number of significant tuples that belong to each aggregate group* and the *error rate* as they directly affect TP-Ag. When the *number of significant tuples that belong to each aggregate group* decreases, this reduces the number of tuples in each sample population. The *smaller the sample population* is the more likely that the result produced may be skewed. Consider a significant tuple t_i that expires before reaching the aggregate operator. Sample population $spop_m$ is the sample population that tuple t_i would have belonged to if tuple t_i had not expired. The aggregate result produced by sample population $spop_m$ will be more affected if the sample population $spop_m$ contains few tuples (smaller population) rather than many tuples (larger population). Decreasing the *error rate* increases the accuracy in the estimated required sample size. This should increase the percentage of correct aggregate results produced by TP-Ag. We varied these variables as they affect TP-Ag's ability to produce accurate results.

7.3   Experimental Findings

Effectiveness at Increasing the Percentage of Correct Aggregate Results Produced. First, we compare the percentage of correct aggregate results produced by each approach. This experiment uses DS1. Figure 3 a shows the average difference between the number of the correct and incorrect aggregate results produced at each minute. This measures whether more correct (positive number) or incorrect results were produced (negative number). Overall TP w/ TP-Ag compared to sem, rand, and trad consistently produces more correct aggregate results.

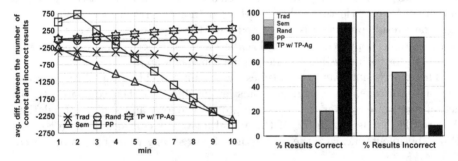

a) Average difference between the number of correct and incorrect aggregate results produced

b) Average % Correct Significant Results Over 10 Min

Fig. 3. Effectiveness at Increasing the % of Correct Aggregate Results

PP produced more correct aggregate results than TP w/ TP-Ag at startup (minutes 1 through 3). However, after the system start-up (minutes 4 through 10) PP produced more incorrect than correct aggregate results. This is as expected. Namely, PP has less overhead than TP w/ TP-Ag. In addition, the aggregate results produced by PP will only be incorrect when the system is overloaded and many significant tuples fail to reach the aggregate operator.

As seen in Figure 3 b shows, compared to all alternative solutions, TP w/ TP-Ag produced a much higher percentage of correct aggregate results. Of all the aggregate results produced by TP w/ TP-Ag, 91.5% were correct. Our results support that TP w/ TP-Ag is effective at increasing the percentage of correct aggregate results produced compared to competitor solutions. ,

TP w/ TP-Ag Versus State-of-the-art Reliable Aggregation Operators for TPs. We now compare TP w/ TP-Ag to state-of-the-art aggregate operators designed to produce reliable results in TPs [4,4,39] (Sec. 1.5). These systems limit which tuples are dropped from specific windows. We refer to these systems as *Shed Window Ag*. First, we compare the percentage of correct aggregate results produced by each approach. This experiment also uses DS1.

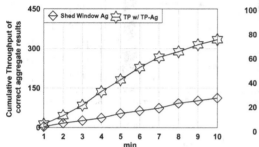
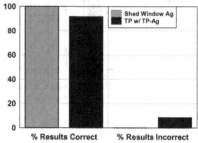

a) Cumulative Throughput of Correct Ag Results   b) Average % Correct Significant Results
Over 10 Min

Fig. 4. TP w/ TP-Ag versus State-of-the-art

As the overall percentage of correct and incorrect significant results in Figure 4 b shows, all aggregate results produced by Shed Window Ag were correct. While, of the aggregate results produced by TP w/ TP-Ag produced 91.5% were correct. Clearly, Shed Window Ag will always produce correct aggregate results. Recall that Shed Window Ag will ensure that no tuples from specific windows are dropped or expire. As a result, Shed Window Ag will only produce correct aggregate results. In contrast, TP w/ TP-Ag seeks to produce results that are estimated to be correct from incomplete windows of tuples.

However, Shed Window Ag may not produce as many aggregate results as TP w/ TP-Ag. As Figure 4 a shows, TP w/ TP-Ag produced roughly 2.9 fold more correct aggregate results than Shed Window Ag. Shed Window Ag will process all tuples (both significant and insignificant) from selected windows. This requires a significant amount of CPU overhead. Hence, Shed Window Ag will not produce as many correct aggregate results as TP w/ TP-Ag.

Clearly, Shed Window Ag and TP w/ TP-Ag have different goals. The goal of Shed Window Ag is to produce correct aggregate results by adjusting how resources are allocated. The goal of TP w/ TP-Ag is to build reliable aggregate results from the significant tuples pulled forward by the TP. Thus, henceforth we no longer compare TP w/ TP-Ag to Shed Window Ag.

Varying the Sample Population Size. We now explore how the number of the significant tuples in each aggregate group affects TP-Ag. All significant tuples belong to two GICS sector groups. This experiment uses four Data Sets (i.e., DS25, DS50, DS75, and DS100). Each Data Set adapts the percentage of significant tuples that belong to the two GICS sector groups. In DS25, 25% of the stocks in the two sectors are significant (75% of these tuples are insignificant). Similarly, in DS50, DS75, and DS100, respectively 50%, 75% and 100% of the tuples in the two sectors are significant. The sample population size increases from DS25 to DS100.

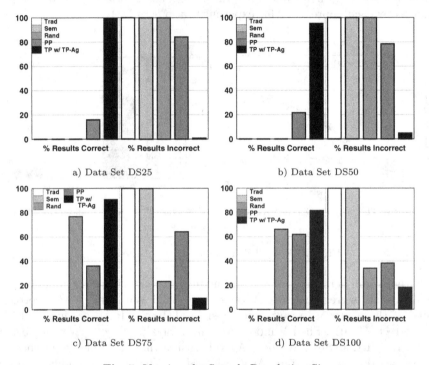

Fig. 5. Varying the Sample Population Size

Figures 5a-5d show the overall percentage of correct and incorrect aggregate results respectively for DS25, DS50, DS75, and DS100. As can be seen, compared to the alternative solutions, TP w/ TP-Ag produced the highest percentage of correct aggregate results. The closest competitors were rand and PP. In DS25 (the smallest sample populations), TP w/ TP-Ag produced 100% and 84.0% more correct aggregate results than rand and PP. In DS100 (the largest sample populations), TP w/ TP-Ag produced 19.1% and 24.2% more correct aggregate results than rand and PP. This is as expected. Namely, TP-Ag achieves the highest gains when the fewest tuples in the sample population fail to reach the aggregate operator. When the stream is saturated with significant tuples, more significant tuples are likely to fail to reach the aggregate operator.

Varying Error Rate. Now, we compare the percentage of correct aggregate results produced by TP w/ TP-Ag when the error rate (i.e., the desired level of precision ϵ of Cochran's sample size formula (Sec. 5.2)) varies. This experiment also uses DS1. We vary the error rate ϵ from 5%, 10%, to 20%. Figure 6 shows the percentage of correct and incorrect aggregate results produced.

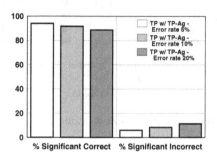

Fig. 6. Varying the error rate

Overall the highest percentage of correct aggregate results was produced when the error rate ϵ is 5%. While the lowest percentage was produced when the error rate ϵ was 20%. The percentage of correct aggregate results produced by TP w/ TP-Ag for the error rate ϵ from 5%, 10%, to 20% was respectively 93.9%, 91.5%, and 88.6%. As expected, decreasing the error rate (i.e., higher level of precision of Cochran's sample size formula) increases the percentage of correct aggregate results achieved by TP w/ TP-Ag (vice versa).

Execution-Runtime CPU Overhead. To measure the runtime overhead we evaluate the cumulative throughput using the worst case scenario for TP w/ TP-Ag. In the worst case scenario, there is adequate resources to process all tuples (Fig. 7 c). As a consequence, for each aggregate result is produced from a sample population that contains all tuples in a window and aggregate group. The overhead of TP systems is the cost to gather and evaluate runtime statistics. In addition, TP-Ag has the additional overhead of tracking statistics to estimate the actual population, evaluating the required sample size (Sec. 5.2), and determining if there is a sample population for each group and window whose size is comparable to the required sample size (Sec. 5.3). This experiment uses DS1.

As can be seen in our results, the difference between the throughput of TP w/ TP-Ag and trad, sem, rand, and PP is respectively 40.2%, 37.9%, 39.1%, and 39.0%. For systems with extremely limited resources, TP w/ TP-Ag may not be a good approach. However, TP w/ TP-Ag is a great fit for systems that require a TP system and desire reliable accuracy in the aggregate results produced.

Memory Overhead. To measure the memory overhead we evaluated the average number of tuples in the state and input queue of the aggregate operator using the worst case scenario outline above (Fig. 7 a & b). As our results demonstrate, the memory overhead of TP w/ TP-Ag is higher than the current state-of-the-art approaches. The state of the aggregate operators in trad, sem, rand, and PP

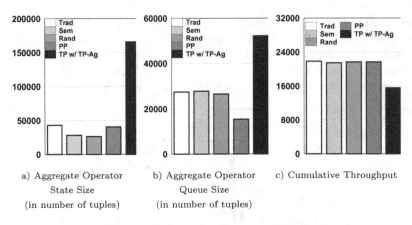

a) Aggregate Operator
State Size
(in number of tuples)

b) Aggregate Operator
Queue Size
(in number of tuples)

c) Cumulative Throughput

Fig. 7. Memory & Execution-Runtime CPU Overhead

respectively have 74.0%, 83.0%, 84.2%, and 75.7% less tuples in their states than TP w/ TP-Ag. While the queues of the aggregate operators in trad, sem, rand, and PP respectively have 47.4%, 46.6%, 49.2%, and 70.6% less tuples in their queues than TP w/ TP-Ag.

This is as expected. Namely, the TP-Ag design relies upon a memory-intensive physical design to support the production of results from subsets of the actual sample population. Again, TP w/TP-Ag is a great fit for systems that require a TP system and desire reliable accuracy in the aggregate results produced. Ensuring the production of reliable aggregate results however carries an overhead.

7.4 Summary of Experimental Findings

We now summarize our key findings.

– TP-Ag is effective at increasing the percentage of correct aggregate results produced in TPs (TP-Ag produces up to 91% more correct aggregate results).
– Decreasing the error rate increases the percentage of correct aggregate results achieved by TP w/ TP-Ag and vice versa.
– TP-Ag is best suited for environments where the stream is not saturated with significant tuples. When the stream is saturated with significant tuples, more significant tuples are likely to expire.
– TP w/TP-Ag is a great fit for systems that require a targeted prioritized data stream system and desire reliable accuracy in the aggregate results produced.

8   Related Work

Below are related works beyond those already covered in Sections 1.5 and 3.

8.1 Aggregate Operators that Support Tuple Level Resource Reduction and Reorder Systems

Some *aggregation operators* proposed to support data stream systems that utilize tuple level resource allocation and reduction aim to only produce non-skewed aggregate results (Sec. 1.5) by requiring that certain tuples from selective windows are never shed. This is limiting in what tuples will and will not be processed. It does not address the TP systems where the user selects which tuples will and will not be processed. These approaches simplify aggregation because they force a complete set of tuples from these windows to arrive at the aggregate operator.

Hellerstein et al. [16] proposed an online interface that allows users to both observe the progress and halt the execution of their aggregation queries. In their approach, load shedding is initiated by the end user. To help ensure the most accurate aggregate results are produced, their approach returns the output in random order, adjusts the rate at which different aggregates are computed, and computes running confidence intervals. The running confidence intervals are displayed to the user.

Babcock et al. [4] proposed a system that supports load shedding. The goal of the system is drop tuples such that accuracy of the aggregate results produced are within certain limits. They consider the probability that dropping certain tuples has on the accuracy of query answers produced by the multiple queries.

Longbo et al. [27] propose a load shedding system for continuous sliding window join-aggregation queries over data streams. Their load shedding strategy partitions the domain of the join attribute into certain sub-domains. Then they filter out selected input tuples based on their join values.

Guo et al. [15] proposed a load shedding approach for aggregation queries with sliding windows. They analyzed the characteristics of subset model and deficiencies of current load shedding methods. Their load shedding algorithm is based on the strategy of dropping tuples from certain window.

Senthamilarasu et al. [35] proposed load shedding techniques for queries consisting of one or more aggregate operators with sliding windows. Their load shedding method utilizes a window function that divides the input into portions of the windows of the aggregate operators. It then utilizes this function to probabilistically determine which tuple to shed.

Akin to these approaches, TP-Ag must also contend with the time versus accuracy trade off. TPs require an aggregate operator that can creates a reliable aggregate result using only the available tuples within the group population. The approach should not adjust how the TP system is allocating resources. It should not change which tuples are pulled forward.

8.2 Tuple Level Resource Reduction

There are many resource allocation approaches that *reduce the workload.* One approach is load shedding. Load shedding drops less significant tuples. It only allocates resources to the tuples not dropped. Once a tuple is chosen to be processed, it will not be shed at any point along the query pipeline.

Aurora [1,45] is a system to manage data streams for monitoring applications. It supports real-time requirements. To achieve this, they proposed using load shedding to reduce the system of less critical tuples. Their key idea was to propose load shedding as a means to control the workload.

Tatbul et al. [38] explored a technique for dynamically inserting and removing drop operators into query plans as required by the current load. They considered both semantic and random shedding. Their cost model does not consider the cost of the drop operators to evaluate tuples. It assumes that this cost is low.

Reiss et al. [33] proposed the Data Triage architecture. It supports systems with bursty arrival rates that can fluctuate. During such bursts, Data Triage captures an estimate of the query results that the system did not have time to compute. They combined these results with the query results to generate more accurate statistics. These statistics are used to evaluate which tuples should be shed.

Tatbul et al. [37] proposed load shedding techniques for distributed stream processing environments. They modeled the distributed load shedding problem as a linear optimization problem. They proposed a distributed approach. It was built for dynamic environments in large-scale deployments.

Nehme et al. [29] proposed a load shedding technique for spatio-temporal stream data. Their load shedding model considered spatio-temporal properties by grouping similarly moving objects into clusters. Then they shed selective objects within each cluster. The locations of the objects shed are approximated based upon their associated clusters.

Wang et al. [41] proposed a load shedding technique for real-time data stream applications. The goal of their approach is to reduce the workload while at the same time preserving the system timing constraints. They proposed different modes. These modes define how the load on the stream is adjusted.

Ma et al. [28] proposed a semantic load shedding technique for real-time data stream applications that utilizes a priority table. It considers both the execution costs and tuple attribute values when deciding which tuples are shed.

Basaran et al. [5] proposed a load shedding method that applies distributed fuzzy logic. It considers the per-stream backlog and selectivity of each query operator. Their approach is event-driven. This allows it to react to bursty workloads.

Lin et al. [25] proposed a linear programming based load shedding method for distributed data stream processing systems. It models the system load as a simple query network with network constraints. It considers two factors. These factors are the amount of available CPU and network resources.

Labrinidis et al. [40] proposed a load shedding strategy that manages the load shedding without requiring any input from users, namely, any manually tuned parameters. Their approach works with complex query networks containing joins, aggregations or shared operators.

In contrast to these approaches, TP seeks to adaptively adjust how resource allocation throughout the query pipeline. These approaches simply decide to process a tuple or not and never revisit this decision. In TP, a tuple may be allocated resources for a portion of the query pipeline. Later on, if more significant tuples are present then this same tuple may be denied resources. This allows the more significant tuples to be processed.

9 Conclusions

This paper makes the following important contributions. Our *TP-Ag* operator tackles the open problem of generating reliable average calculations for normally distributed data from incomplete aggregation populations resulting from decisions made by TPs. TP-Ag produces non-skewed average calculations by determining at run-time which combination of subset(s) of an aggregation population (if any) are used to generate a result. A carefully designed application of Cochran's sample size methodology is used to measure the accuracy of possible populations. Our experimental study confirms that TP-Ag is effective at increasing the percentage of reliable results produced in TPs (TP-Ag produces up to 91% more accurate results).

Acknowledgments. We thank our WPI peers for CAPE [34] and feedback. We also thank GAANN and NSF grants: IIS-1018443, 0917017, 0414567, and 0551584 for their support.

References

1. Abadi, D.J., Carney, D., Çetintemel, U., Cherniack, M., Convey, C., Lee, S., Stonebraker, M., Tatbul, N., Zdonik, S.: Aurora: A new model and architecture for data stream management. The International Journal on Very Large Data Bases, 120–139 (2003)
2. Abadi, D.J., et al.: Aurora: A new model and architecture for data stream management. VLDB Journal, 120–139 (2003)
3. Arasu, A., et al.: The cql continuous query language: semantic foundations and query execution. VLDB Journal, 121–142 (2006)
4. Babcock, B., et al.: Load shedding for aggregation queries over data streams. In: ICDE, p. 350 (2004)
5. Basaran, C., Kang, K.-D., Zhou, Y., Suzer, M.H.: Adaptive load shedding via fuzzy control in data stream management systems. In: 2012 5th IEEE International Conference on Service-Oriented Computing and Applications (SOCA), pp. 1–8. IEEE (2012)
6. Carney, D., et al.: Monitoring streams: A new class of data management applications. In: VLDB, pp. 215–226 (2002)
7. Cochran, W.G.: Sampling Techniques, 3 edn. John Wiley (1977)
8. Cormode, G., Korn, F., Tirthapura, S.: Time-decaying aggregates in out-of-order streams. PODS, 89–98 (2008)
9. Das, A., et al.: Semantic approximation of data stream joins. IEEE, 44–59 (2005)
10. Dobra, A., et al.: Processing complex aggregate queries over data streams. In: SIGMOD, pp. 61–72 (2002)
11. Fama, E.F.: The behavior of stock-market prices. The Journal of Business **38**(1), 34–105 (1965)

12. Finance, Y.: http://finance.yahoo.com/
13. Gainey, R.R., et al.: Understanding the experience of house arrest with electronic monitoring: An analysis of quantitative and qualitative data. International Journal of Offender Therapy and Comparative Criminology (2000)
14. Golab, L., et al.: Update-pattern-aware modeling and processing of cont. queries. In: SIGMOD, pp. 658–669 (2005)
15. Guo, J.-F., He, C.-L.: Load shedding for sliding window aggregation queries over data streams. Application Research of Computers, 1–23 (2009)
16. Hellerstein, J.M., Haas, P.J., Wang, H.J.: Online aggregation. SIGMOD **26**(2), 171–182 (1997)
17. Hoeffding, W.: Probability Inequalities for Sums of Bounded Random Variables. Journal of the American Statistical Association **58**(301), 13–30 (1963)
18. Hoyle, S.: Use and abuse of statistics. ASLIB Proc. **40**(11–12), 321–324 (1988)
19. Kang, H.G., Mahoney, D.F., Hoenig, H., Hirth, V.A., Bonato, P., Hajjar, I., Lipsitz, L.A.: In situ monitoring of health in older adults: technologies and issues. Journal of the American Geriatrics Society **58**(8), 1579–1586 (2010)
20. Kargupta, H., Park, B.-H., Pittie, S., Liu, L., Kushraj, D., Sarkar, K.: Mobimine: monitoring the stock market from a pda. SIGKDD Explor. Newsl. **3**(2), 37–46 (2002)
21. Katopodis, P., et al.: A hybrid, large-scale wireless sensor network for missile defense. IEEE, 1–5 (2007)
22. Li, J., et al.: No pane, no gain: efficient evaluation of sliding-window aggregates over data streams. SIGMOD **34**, 39–44 (2005)
23. Li, J., et al.: Semantics and evaluation techniques for window aggregates in data streams. SIGMOD, 311–322 (2005)
24. Lin, C.-C., et al.: Wireless health care service system for elderly with dementia. IEEE, 696–704 (2006)
25. Lin, O., Qin, Z., Jingjing, Q., Qiumei, P.: A new linear programming based load-shedding strategy. In: 2012 11th International Symposium on Distributed Computing and Applications to Business, Engineering & Science (DCABES), pp. 260–263. IEEE (2012)
26. Liu, B., et al.: Run-time operator state spilling for memory intensive long-running queries. SIGMOD, 347–358 (2006)
27. Longbo, Z., Zhanhuai, L., Zhenyou, W., Min, Y.: Semantic load shedding for sliding window join-aggregation queries over data streams. In: International Conference on Convergence Information Technology, pp. 2152–2155 (2007)
28. Ma, L., Zhang, Q., Shi, N.: A semantic load shedding algorithm based on priority table in data stream system. In: International Conference on Fuzzy Systems and Knowledge Discovery, pp. 1167–1172 (2010)
29. Nehme, R.V., Rundensteiner, E.A.: Clustersheddy: Load shedding using moving clusters over spatio-temporal data streams. In: Kotagiri, R., Radha Krishna, P., Mohania, M., Nantajeewarawat, E. (eds.) DASFAA 2007. LNCS, vol. 4443, pp. 637–651. Springer, Heidelberg (2007)
30. Network, M.: Where have all the investors gone? (February 2012). http://money.msn.com
31. Olston, C., Widom, J.: Offering a precision-performance tradeoff for aggregation queries over replicated data. Technical Report 2000–16, Stanford InfoLab (2000)
32. Press, A.: Officials lose track of 16,000 sex offenders after gps fails (2010). http://www.foxnews.com

33. Reiss, F., Hellerstein, J.M.: Data triage: An adaptive architecture for load shedding in telegraphcq. In: IEEE International Conference on Data Engineering, pp. 155–156 (2005)

34. Rundensteiner, E.A., et al.: Cape: Continuous query engine with heterogeneous-grained adaptivity. In: VLDB, pp. 1353–1356 (2004)

35. Senthamilarasu, S., Hemalatha, M.: Load shedding techniques based on windows in data stream systems. In: 2012 International Conference on Emerging Trends in Science, Engineering and Technology (INCOSET), pp. 68–73. IEEE (2012)

36. Tatbul, N.: QoS-driven load shedding on data streams. In: Chaudhri, A.B., Unland, R., Djeraba, C., Lindner, W. (eds.) EDBT 2002. LNCS, vol. 2490, pp. 566–576. Springer, Heidelberg (2002)

37. Tatbul, N., Çetintemel, U., Zdonik, S.: Staying fit: Efficient load shedding techniques for distributed stream processing. In: International Conference on Very Large Data Bases, pp. 159–170 (2007)

38. Tatbul, N., et al.: Load shedding in a data stream manager. In: VLDB, pp. 309–320 (2003)

39. Tatbul, N., Zdonik, S.: Window-aware load shedding for aggregation queries over data streams. VLDB, 799–810 (2006)

40. Pham, T.N., Chrysanthis, P.K., Labrinidis, A.: Self-managing load shedding for data stream management systems, 1–7 (2013)

41. Wang, H.-Y., Qin, Z.-D., Li, B.-Y., Cong, J., Wang, Z.-J., Du, M.: Novel load shedding approach for real-time data stream processing. Journal of Chinese Computer Systems, 1–4 (2010)

42. Wei, M., et al.: Achieving high output quality under limited resources through structure-based spilling in xml streams. PVLDB, 1267–1278 (2010)

43. Works, K., Rundensteiner, E.: Preferential resource allocation in stream processing systems. International Journal of Cooperative Information Systems (2014)

44. Works, K., Rundensteiner, E.A.: The proactive promotion engine. In: ICDE, pp. 1340–1343 (2011)

45. Zdonik, S.B., et al.: The aurora and medusa projects. IEEE, 3–10 (2003)

Slicing the Dimensionality: Top-k Query Processing for High-Dimensional Spaces

Gheorghi Guzun$^{(\boxtimes)}$, Joel Tosado, and Guadalupe Canahuate

Department of Electrical and Computer Engineering,
The University of Iowa, Iowa City, IA, USA
{gheorghi-guzun,joel-tosadojimenez,guadalupe-canahuate}@uiowa.edu
http://www.uiowa.edu

Abstract. Top-k (preference) queries are used in several domains to retrieve the set of k tuples that more closely match a given query. For high-dimensional spaces, evaluation of top-k queries is expensive, as data and space partitioning indices perform worse than sequential scan. An alternative approach is the use of sorted lists to speed up query evaluation. This approach extends performance gains when compared to sequential scan to about ten dimensions. However, data-sets for which preference queries are considered, often are high-dimensional. In this paper, we explore the the use of bit-sliced indices (BSI) to encode the attributes or score lists and perform top-k queries over high-dimensional data using bit-wise operations. Our approach does not require sorting or random access to the index. Additionally, bit-sliced indices require less space than other type of indices. The size of the bit-sliced index (without using compression) for a normalized data-set with 3 decimals is 60 times smaller than the size of sorted lists. Furthermore, our experimental evaluation shows that the use of BSI for top-k query processing is more efficient than Sequential Scan for high-dimensional data. When compared to Sequential Top-k Algorithm (STA), BSI is one order of magnitude faster.

Keywords: Top-k queries · Preference queries · High-dimensional data

1   Introduction

Top-k processing techniques attempt to efficiently find the top k results from data sources. Objects, or data items, pertaining to these data sources may be described through multiple numerically-valued attributes, or dimensions. An object's numeric value for a specific attribute is its local score for that attribute. Top-k techniques rely on ranking functions in order to determine an overall score for each of the objects across all the relevant attributes being examined [1]. This overall score may then be used to choose the top-k results. Top-k processing is a crucial requirement across several applications or domains. In the multimedia domain numerically-valued feature vectors describe the multimedia objects. An image, for example, may be described by thousands of feature vectors [2–4].

© Springer-Verlag Berlin Heidelberg 2014
A. Hameurlain et al. (Eds.): TLDKS XIV, LNCS 8800, pp. 26–50, 2014.
DOI: 10.1007/978-3-662-45714-6_2

Users usually search for multimedia objects for which they only desire the best matching result that derive from the overall grade of match, or overall score of the features. [5,6].

In the domain of information retrieval, consider a search engine tasked in retrieving the top-k results from various sources. In this scenario, the ranking function considers scores based on word-based measurements, as well as hyperlink analysis, traffic analysis, user feedback data, among others, to formulate its top-k result [7,8].

Other domains and applications such as monitoring networks [9,10], P2P systems [11], data stream management streams [12], restaurant selection systems [13], among others, also rely on top-k processing techniques [14,15].

Data in these domains is often high-dimensional with over a hundred of attributes. The curse of dimensionality relates to the negative impact the increase of dimensionality has on query processing techniques. Top-k techniques express their effects through either high time or space complexity as dimensionality increases [16]. E.g. indexing structures are not efficient for dimensionality sometimes as low as 6 dimensions[17]. Moreover, this effect pervades to the extent where the top-k techniques perform worse than sequential scan [17].

A popular top-k processing technique, the Threshold Algorithm (TA) [14], and some of its optimizations [17–21] involve the use of sorted lists for each attribute. The idea is that the collection of these sorted lists allow an efficient computation of the top-k aggregated overall scores with as few accesses to the data as possible. These sorted lists avoid scanning of the entire list of entries for each of the attributes under consideration. Moreover, a threshold value is used to determine when the algorithm should stop while guaranteeing the top-k results. TA-based techniques extend the performance benefits to tens of dimensions.

In this paper, we explore the evaluation of top-k queries in high-dimensional spaces without the use of sorted lists or hierarchical indices. We propose the use of bit-sliced indices (BSI) to encode the score list and perform top-k queries over these high-dimensional data. Since our index is not sorted, new data is just appended to the end. The bit-slices are added and resulting in another BSI. From this aggregate ranking a top-k result can be derived using bit-wise operations.

The primary contributions of this paper can be summarized as follows:

- We developed efficient algorithms to evaluate top-k queries over bit-sliced indexing exclusively using bit-wise operations.
- We analyze the cost of the proposed algorithms in terms of algorithm complexity.
- We evaluate three types of top-k queries: *top-k queries, Boolean preference queries, weighted preference queries.*
- We evaluate the cost of the proposed approach in terms of index size and also query time.
- We compare performance gains against Sequential Scan (SS), Threshold Algorithm (TA), Sequential Top-k Algorithm(STA) and Best Position Algorithms (BPA and BPA2).
- We perform experiments over both synthetic and real datasets.

Table 1. Notation Reference

Notation	Description
n	Number of rows in the data
m	Number of attributes in the data
s, p	Number of slices used to represent an attribute
w	Computer architecture word size
Q	Query vector
$|q|$	Number of non-zero preferences in the query
b	Number of bits used to represent a query preference

The rest of the paper is organized as follows. Section 2 presents background and related work. Section 3 describes the problem formulation and the proposed solution using bit-sliced indices. Section 4 presents the cost analysis of the proposed algorithms. Section 5 shows experimental results. Finally, conclusions are presented in Section 6.

2 Background and Related Work

This section presents background information for bit-sliced indices and related work for top-k query processing. For clarity, we define the notations used further in this paper in Table 1.

2.1 Bit-Sliced Indexing

BSI [22,23] can be considered a special case of the encoded bitmaps [24]. With bit-sliced indexing, binary vectors are used to encode the binary representation of the attribute value. Only $\lceil \log_2 values \rceil$ vectors are needed to represent all values. One BSI is created for each attribute.

Figure 1 shows an example of the BSIs for two attributes and their sum. Since each attribute has three possible values, the number of bit-slices for each BSI is 2. For the sum of the two attributes, the maximum value is 6, and the number of bit-slices is 3. The first tuple t_1 has the value 1 for attribute 1, therefore only the bit-slice corresponding to the least significant bit, $B_1[0]$ is set. For attribute 2, since the value is 3, the bit is set in both BSIs. The addition of the BSIs representing the two attributes is done using efficient bit-wise operations. First, the bit-slice $sum[0]$ is obtained by XORing $B_1[0]$ and $B_2[0]$ i.e. $sum[0] = B_1[0] \oplus B_2[0]$. Then $sum[1]$ is obtained in the following way $sum[1] = B_1[1] \oplus B_2[1] \oplus (B_1[0] \wedge B_2[0])$. Finally $sum[2] = Majority(B_1[1], B_2[1], (B_1[0] \wedge B_2[0]))$.

BSI arithmetic for a number of operations is defined in [23]. We adapt two of these bit-sliced operations (sum and topK) as components in our top-k query processing and define a new operation to multiply a BSI by the query preference.

	Raw Data		Bit-Sliced Index (BSI)				BSI SUM		
			Attrib 1		Attrib 2				
Tuple	Attrib 1	Attrib 2	$B_1[1]$	$B_1[0]$	$B_2[1]$	$B_2[0]$	$sum[2]^3$	$sum[1]^2$	$sum[0]^1$
t_1	1	3	0	1	1	1	1	0	0
t_2	2	1	1	0	1	0	0	1	1
t_3	1	1	0	1	0	1	0	1	0
t_4	3	3	1	1	1	1	1	1	0
t_5	2	2	0	1	1	0	1	0	0
t_6	3	1	1	1	0	1	1	0	0

$^1 sum[0] = B_1[0]$ XOR $B_2[0]$, $C_0 = B_1[0]$ AND $B_2[0]$
$^2 sum[1] = B_1[1]$ XOR $B_2[1]$ XOR (C_0)
$^3 sum[2] = C_1 = Majority(B_1[1], B_2[1], (C_0))$

Fig. 1. Simple BSI example for a table with two attributes and three values per attribute

2.2   Top-k Queries

Numerous techniques have been proposed to process top-k queries using a variety of data and space partitioning indices. However, their performance degrades for high-dimensional spaces [1,17,25].

The preferred algorithms for multi-dimensional spaces stem from the Threshold Algorithm (TA). Thus, before addressing these optimizations, here we present the TA algorithm [14].

1. Do sorted access in parallel to each of the m sorted lists L_i. As an object R is seen under sorted access in some list, do random access to the other lists to find the grade x_i of object R in every list L_i. Then compute the grade $t(R) = t(x_1, \ldots, x_m)$ of object R. If this grade is one of the k highest we have seen, then remember object R and its grade $t(R)$ (ties are broken arbitrarily, so that only k objects and their grades need to be remembered at any time).
2. For each list L_i, let x_i^L be the grade of the last object seen under sorted access. Define the threshold value τ to be $t(x_1^L, \ldots, x_m^L)$. As soon as at least k objects have been seen whose grade is at least equal to τ, then halt.
3. Return the top-k objects as the query answer.

Correctness of TA follows from the fact that the threshold value τ represents the best possible score that any object not yet seen can have, and TA stops when it can guarantee that no unseen object might have a better score than the current top-k ones [1,21,26].

The Best Position Algorithms (BPA and BPA2) [15], were proposed as optimizations of TA where the threshold condition is improved. It is determined from the best positions from each of the lists above which all scores have been seen. BPA2 further improves over BPA by avoiding accessing the same position several times, by doing direct access to the position which is just after the best position. The IO-Top-k technique [27] involves a balancing of seek time and transfer rate by mixing the amount of sorted and random accesses performed. Moreover, random accesses are controlled given the probability of a certain record to be in the top-k.

While TA relies on sorted and random access, the NRA variant [14] has no random access for applications where random access is overly expensive. The approach consists of two phases, one where it extracts candidate objects gradually from the top of the sorted lists. This will continue until the threshold condition is met. The second phase will gradually establish that no remaining candidates are better than the current top-k and will then terminate. The Threshold Algorithm over Bucketized Sorted Lists with Bloom Filters (TBB) [26] improves upon NRA by estimating the sequential scanning required and thus reducing disk access. The 3-phased no random access algorithm (3P-NRA) also improves upon NRA [28] by reducing the expensive probing of candidates in the NRA algorithm. However, 3P-NRA and TBB generally incur the large maintenance cost of the potential candidates and sorted lists [21].

Numerous other techniques have been developed for the top-k query processing that also originate, or involve concepts, from the TA technique [18–20,29,30]. However, these and other top-k techniques such as view-based techniques do not perform well for high dimensional data [1,17,21].

Recently, Sequential Top-k Algorithm (STA) was proposed to preprocess the sorted lists in order to reduce the space complexity and allow for the early termination feature, by a threshold value, as in TA. It eliminates the need to store the object identifier for each attribute as is the case when using sorted lists. Instead it stores a single identifier with all the attributes and a leaner indicator attribute (1 byte to address 256 attributes). This indicator attribute determines when an object is first seen, essentially capturing when a new object is accessed as in the sorted access of TA. Additionally, the resulting table from the preprocessing is sequentially accessed and used to dynamically update the threshold element. Furthermore, STA also avoids random access, such as in NRA, when evaluating top-k queries. In its paper, STA was compared against TA, 3P-NRA, BPA, IO-Top-k, and TBB. The experimental results on synthetic data sets for increasing dimensionality (up to 16) state a 1-2 order of magnitude improvement over these other top-k processing techniques [21]. Thus, we compare STA against our approach.

Note that our approach does not maintain sorted lists nor does it use a threshold computation as a stopping condition. Bit-Sliced Indexing (BSI) is used to compute the scores and then process the top-K query using fast logical operations supported by hardware. Since the scores are computed for all the tuples, performance is not considerably affected by the value of k. This is in contrast to both sequential scan and other list-based approaches (TA, STA, etc.) which maintain an auxiliary data structure of size $O(k)$. Additionally, the BSI index being a variant of the bitmap index, can benefit from word-aligned compression and can efficiently support other types of queries such as selection and aggregation queries.

3 Proposed Approach

In this section we first formulate the top-k queries supported in this paper and then describe the query execution algorithm using bit-slice indexing.

3.1 Problem Formulation

Consider a relation R with m attributes or numeric scores and a preference query vector $Q = \{q_1, \ldots, q_m\}$ with m values where $0 \leq q_i \leq 1$. Each data item or tuple t in R has numeric scores $\{f_1(t), \ldots, f_m(t)\}$ assigned by numeric component scoring functions $\{f_1, \ldots, f_m\}$. The combined score of t is $F(t) = E(q_1 \times f_1(t), \ldots, q_m \times f_m(t))$ where E is a numeric-valued expression. F is monotone if $E(x_1, \ldots, x_m) \leq E(y_1, \ldots, y_m)$ whenever $x_i \leq y_i$ for all i. In this paper we consider E to be the summation function: $F(t) = \sum_{i=1}^{m} q_i \times f_i(t)$. The k data items whose overall scores are the highest among all data items, are called the top-k data items. We refer to the definition above as **top-k weighted preference query**.

In this paper we also consider two special cases of query vectors:

- **top-k baseline**: the query vector is all 1s.
- **top-k boolean**: the query vector is either 0 or 1 for each attribute.

These two specializations of the top-k preference query are less expensive to compute than the general top-k preference query but also have numerous applications in the literature, e.g. spatial searches with text constraints in geographical collections [31].

The rest of this section describes the proposed approach for answering the general top-k preference queries.

3.2 Top-k Preference Query Execution

Let us denote by B_i the bit-sliced index (BSI) over attribute i. A number of slices s is used to represent values from 0 to $2^s - 1$. $B_i[j]$ represents the jth bit in the binary representation of the attribute value and it is a binary vector containing n bits (one for each tuple). The bits are packed into words, the storage requirement for each binary vector is n/w, where w is the computer architecture word size (64 in our implementation).

In order to compute the score for each data point we first multiply the attribute value by the query preference for that attribute using bit-wise operations. Given a query Q, the query vector is first converted to integer weights based on the desired precision. Let us denote by b, the number of bits used to represent a query preference. The preference query execution algorithm pseudocode is given in Algorithm 1.

We identify three main parts in our main algorithm 1:

1. For all non-zero weights, multiply the BSI for the attribute with the corresponding query weight (Lines 3 and 6).

```
prefQuery (B,q,k)

1:   if (k < 0)
2:      Error ("k is invalid")
3:   S = Multiply (q_j, B_j) where j is the first non-zero weight in q
4:   for (i = j + 1; i ≤ m; i++)
5:      if q_i > 0
6:         S = SUM_BSI(S, Multiply (q_i, B_i))
7:   T = TopK (S, k)
8:   return T
```

Algorithm 1. Preference query execution using bit-slices. B is the set of all BSIs, q is the query vector, and k is the desired number of results.

2. Sum the partial scores produced by the Multiply algorithm into a BSI S (Line 6).
3. Find the top k data points given the final BSI score S.

Algorithm 1 calls the *Multiply* function on line 3. Algorithm 2 shows the pseudocode for multiplication of a query weight with the attribute values. It follows the same logic as a sequential multiplier with the difference that the multiplicand is a BSI, not a single number.

First, the three bit-arrays R, S and C are initialized to be all-zeros. R is the product BSI, while S and C represent the sum, and carry bit-vectors for every slice. When the multiplier's least significant bit is set (Line 5), the multiplicand's BSI is shifted and added to the result (Lines 6-13). Otherwise the multiplicand's BSI is shifted until it finds its least significant bit set. If the final bit-array C, representing the carry bits has at least one set bit (Line 14), then the number of slices in the result R is incremented by one (Lines 15-16). The multiplier is shifted before the next iteration of the loop (Line 17). The shift amount is stored as an offset counter (Line 18).

The return BSI R represents the partial score of each tuple for a single attribute. Shifting the bit-slices does not impact the performance cost of the addition operation, as the shifting offset is simply passed along when accessing the bit-slices.

Further, Algorithm 1 calls the *SUM_BSI* function on line 6. Algorithm 3 shows the pseudocode for the addition of two BSIs, A and B, with a number of slices s and p, respectively. The result is another BSI S with MAX$(p, s)+1$ slices. A "Carry" bit-slice C is used in Algorithm 3 whenever two or three bit-slices are added to form S_i, and a non-zero C must then be added to the next bit-slice S_{i+1}. Once the bit-slices in either A or B are exhausted, calculations of C are likely to result in zero in very few iterations soon after. An example of a BSI sum was shown earlier in Figure 1.

Finally, the top k tuples are selected by calling the top-k algorithm presented in Algorithm 4 over the sum BSI representing the scores. The top-k algorithm

---

Multiply (q_i, B)

```
1:   R = ∅                                    //The product BSI
2:   offset=0                                 //Shift factor
3:   bitArray S = ∅,C = ∅                     //Sum and carry bits
4:   while (q_i > 0)
5:       if (number & 1 == 1)                 //if the last bit is set
6:          for (i = 0; i < B.slices; i++)    //Add the slices
7:              S = B[i] XOR R[i+offset] XOR C
8:              C = Majority(B[i],R[i+offset],C)
9:              R[i+offset]=S;
10:         for (j = i+offset; j < B.slices; j++) //Add C to the remaining slices
11:             S = R[j] XOR C;
12:             C = R[j] AND C;
13:             R[j] = S;                      // add the slice S to the product
14:         if (count(C) > 0)
15:             R[B.slices]=C;                 // Carry bit
16:             R.slices++;
17:       q_i >>= 1;                           // shift the query weight
18:       offset++;                            //update the offset
19:  return R;
```

Algorithm 2. Multiplication algorithm. q_i is the integer representation of the query weight for attribute i and B is the BSI representing attribute i.

starts evaluating the scores from the most significant bits. Variables used in Algorithm 4 exist from one loop pass to the next. The bitArrays G, E and X, and positive integer *count* are only temporary, used to hold results within a loop pass for efficiency. The G bit-array represents the data points with larger values seen so far, and the E bit-array represents the data points that are currently tied. We initialize E to be an all 1s bit-array and G as an empty set.

In the first iteration, X gets the values from the most significant bit-slice. If the number of set bits in this slice is greater than k then E is also assigned $S[i]$. Otherwise E will store the tuples that do not have the most significant bit set. For the next iteration X is assigned G OR $(E$ AND $S[i])$ (the tuples in E that also have set bits in $S[i]$). Then, if the number of set bits in X is greater than k $E = E$ AND $S[i]$ otherwise $E = E$ AND $S[i]$. Algorithm 4 will continue iterating through the bit-slices of S until *count* equals to k and will return the resulting k tuples in the form of a bit-vector (F).

Example: Putting It All Together. In this sub-section we explain how Algorithms 1, 2, 3 and 4 work together through a simple example. The numbers in parentheses reference to the numbers in Figure 2.

As an illustrative example, consider a table with two attributes where each tuple represents a bag of white and black pebbles. Pebbles' counts correspond to attributes "White" and "Black". Figure 2 shows an example of a top-k preference

```
SUM_BSI(A,B)

1:    S[0] = A[0] XOR B[0]              // bit on in S[0] iff bit on either A[0] or B[0]
2:    C = A[0] AND B[0]                 // C is "Carry" bit-slice
3:    for (i = 1; i <MIN(s,p); i++)     // While there are bit-slices in A and B
4:        S[i] = (A[i] XOR B[i] XOR C)  // one (or three) bit on gives bit on in S_i
5:        C =Majority(A[i], B[i], C)    // two (or more) bits on gives bit on in C
6:    if (s > p)                        // if A has more bit-slices than B
7:        for (i = p + 1; i ≤ s; i++)   // continue loop until last bit-slice
8:            S[i] = (A[i] XOR C)       // 1 bit on gives bit on in S[i]; C might be 0
9:            C = (A[i] AND C)          // two bits on gives bit on in C
10:   else                             // B has at least as many bit-slices as A
11:       for (i = s + 1; i ≤ p; i++)   // continue loop until last bit-slice
12:           S[i] = (B[i] XOR C)       // one bit on gives bit on in S[i]
13:           C = (B[i] AND C)          // two bits on gives bit on in C
14:   if (C is non-zero)                // if still non-zero Carry after bit-slices end
15:       S[MAX(s,p) + 1] = C           // Put Carry into final bit-slice of S
```

Algorithm 3. Addition of BSIs. Given two BSIs, A and B, we construct a new sum BSI, $S = A + B$, using the following pseudo-code. We must allow the highest order slice of S to be $S[MAX(s,p) + 1]$, so that a carry from the highest bit-slice in A or B will have a place.

query over this table using the Bit-Sliced Index Arithmetic. The query asks for the two bags, out of all of bags, that contain the most pebbles given a preference query with weights 0.4, and 0.6 for the white and black pebbles, respectively.

Initially, the data is indexed using a BSI, and each column in the attribute bitmap is saved as a bit-vector called bit-slice (the most significant bit-slice of attribute "White" is shaded in Figure 2). Since this data has a low value range, there are only three bit-slices per attribute. All the bit-slices form the BSI index. Using the BSI index and Algorithm 1, the query will be performed as illustrated in Figure 2. The weight multiplication is performed by shifting and adding slices, as described in Algorithm 2. The query weights are treated as integers, and thus 0.4 and 0.6 are scaled to 4 and 6 respectively. For every set-bit in the transformed query weight, the BSI index is shifted to the left by the number that describes the set-bit position in the query bit-string ($offset$ in Algorithm 2). For example, the query weight for the first attribute in the figure, has only one set bit. Thus Algorithm 2 "shifts" the BSI for this attribute to the left by two 2 (second rightmost is the position of the set-bit - starting from 0) (1). This result is then added to a sum BSI using algorithm 3 (8). For the second attribute, Algorithm 2 first shifts the BSI by 2 and keeps it as an intermediate result (R in Algorithm 2) (3,4), then shifts it by 1 and adds it to R again (5,6). Once Algorithm 2 completes the multiplication, its result is added to the sum BSI, using Algorithm 3 (7,8).

After the multiplications and additions are complete for all the attributes, Algorithm 4 is applied over the sum BSI (9). Algorithm 4 traverses the sum BSI

TopK(S,k)

1:   $G = \emptyset$
2:   $E =$ All1s   // Initialize an all ones bitVector
3:   **for** $(i = s - 1; i \geq 0; i-)$ //While there are bit-slices in S
4:        $X = G$ OR $(E$ AND $S[i])$ //X is trial set: G OR
                                          //(rows in E with 1-bit in position i)
5:        **if** $((count =$COUNT$(X))> k)$ //If more candidates than k
6:             $E = E$ AND $S[i]$ //Intersection of E and current slice
7:        **else if** $(count < k)$
8:             $G = X$ //G in next pass gets all rows in X
9:             $E = E$ AND NOT$(S[i])$ //E in next pass contains no rows r
                                          //with bit i on in S(r)
10:       **else**   // count==k
11:            $E = E$ AND $S[i]$ //all rows r with bit i on in
                                          //S(r) will be in E
12:            **break**
13: $count =$COUNT(G) //Update count
14: **if** $count < k$ //if count is greater than k
15:       $E = $pick$(E,k - count)$ //Pick k-ncount candidates to break ties
16: **else**
17:       $E = \emptyset$
18: $F = G$ OR E
19: **return** F

Algorithm 4. Find k tuples with the largest values in S (ties are arbitrarily broken), s is the number of slices in S.

starting from the most significant bit (from left to right), and finds the top-k tuples by eliminating candidates. First it eliminates all the tuples that do not have a set-bit in the left-most bit-slice, and saves the tuples that have a set-bit in an intermediate result (X in Algorithm 4). Then it moves to the second left-most bit-slice. In this example the number of tuples that have set-bits and also had set-bits in the previous bit-slice is smaller than two (only one). This single tuple is marked as to be saved in X. Everything else is also kept in G (Algorithm 4). For the next bit-slice, there are two tuples with set-bits, that are also in G. These tuples, plus the one in X are now saved to G. At this point the intermediate result contains 3 tuples. Algorithm 4 will continue this way until the desired number of tuples remains in the intermediate result. In Figure 2 the goal is to extract top-2 tuples, and thus Algorithm 4 will stop once this number is achieved. If the number of slices is exhausted and Algorithm 4 is not able to discriminate between some tuples (i.e. their score is the the same), ties are broken arbitrarily. Finally, the result is returned in form of a bit-slice, having set-bits for the tuples that meet the user's criteria (10).

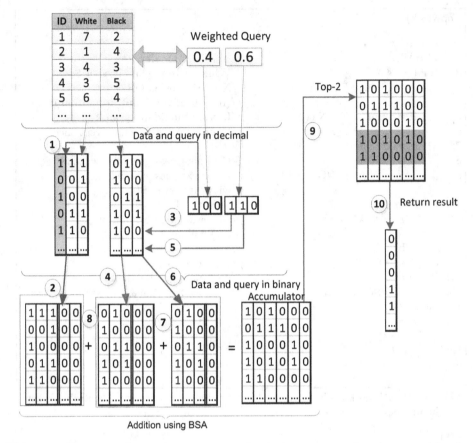

Fig. 2. Example of BSI Arithmetic applied for finding top-2 tuples given a weighted preference query

4   Cost Analysis

In this section we analyze the cost of computing top-k preference queries using the proposed BSI approach both in terms of index size and query execution time.

4.1   Index Size

The analysis is based on verbatim or uncompressed bit-sliced indexing (BSI). The BSI size is independent of data distribution and only depends on the number of rows n and the number of slices used to represent each attribute. Each bit slice has n bits, which are packed into words. The number of slices per attribute can be computed from the attribute cardinality c_i (number of distinct values) as: $s_i = \lceil \log_2 c_i \rceil$. The index size (in bits) can then be computed as:

$$IndexSize = \sum_{i=1}^{m} s_i n \qquad (1)$$

where m is the number of attributes, n is the number of tuples, and s_i is the number of slices used to represent the value of attribute i. For top-k queries it makes sense to normalize the attribute values to avoid one attribute with extremely large values to dominate the final score. Assuming the attributes are normalized between 0 and 1, the same number of slices can be used for all the attributes. The number of slices depends on the desired resolution to represent the normalized attribute values. For approximations using 3, 4, or 6 decimals the number of slices s would be 10, 14, or 20, respectively. In this case, the BSI size formula can be simplified as:

$$IndexSize = m \times s \times n \qquad (2)$$

As long as the number of slices used is less than the computer word size, BSI will always require less storage than the original data. The sorted lists, on the contrary require at least double the storage space than the original data. The reason is that each list stores the attribute value and the tuple id. BSI uses the bit position as the tuple id and since the slices are not sorted, the same position in all the slices always refer to the same tuple.

Note that the number of slices used to represent the attributes does not bound the number of slices to represent the BSI summation of these attributes. The resulting sum BSI is expected to have more slices, and since slices are added as bitVectors, there is no overflow in the computation.

4.2   Query Execution

The top-k preference queries are processed using bit-wise operations over the bit-slices. If the bit operations are performed on a computer with a 64-bit architecture, then the number of bit operations that we need to perform to operate two bitmaps is given by $n/64$. In general, with a word size equal to w-bits, each individual bit-wise operation would process w tuples simultaneously. The total cost of the query execution using bit-sliced indices is determined by the number of bitmaps that are operated together and the number of bit-wise operations performed.

Top-k Weighted Preference Queries. Algorithm 1 performs the generic top-k weighted preference query. The total cost of the preference query processing defined in Algorithm 1 is given by:

$$O\left(|q|bs\frac{n}{w}\right) \qquad (3)$$

where $|q|$ is the number of non-zero preferences in the query (m in the worst case), b is the number of bits used to represent the preferences for each attribute query weight, s is the number of slices used to represent each attribute, and n is the number of tuples in the data set.

Algorithm 1 calls the Multiply Algorithm $|q|$ times, the SUM_BSI Algorithm $|q| - 1$ times, and the top-k Algorithm once.

The cost of Algorithm 2 in the worst case can be expressed in terms of BSI bit-wise operations $O(bs)$, where b is the number of bits used to represent the query preference and s is the number of slices used to represent each attribute. To represent preferences with 1 or 3 decimals, the number of slices would be 4 or 10, respectively. We can expect the number of set bits in each preference to be half the bits (on average). The cost of each bit-wise operation is $O(n/w)$, which allow us to express the cost of the multiply algorithm as:

$$O\left(bs\frac{n}{w}\right) \tag{4}$$

The number of slices in the result R is at most $s + p$ slices. Note that the number of slices per attribute depends on the value range of the attribute over the entire data set. In the cases where less slices are needed than the number of bits required to store the attribute data type, then our index is guaranteed to be smaller than the raw data. This is the case for most real data sets, which means that the BSI index size is usually smaller than the raw data even without applying compression.

The cost of Algorithm 3 is dependent on the maximum number of slices between A and B. The number of slices in B is (in the worst case) $p + s$, while the number of slices in A (the partial score computed so far) is $p + s + \lceil \log_2 m \rceil$ (in the worst case). The cost of computing the sum BSI can then be expressed as:

$$O\left((b + s + \lceil log_2 m \rceil)\frac{n}{w}\right) \tag{5}$$

Finally, the cost of computing the top-k scores is given by the number of slices in S $(p + s + \lceil \log_2 m \rceil)$:

$$O\left((b + s + \lceil log_2 m \rceil)\frac{n}{w}\right) \tag{6}$$

Equation 3 represents the upper bound for equations 4, 5, 6, and thus gives the upper bound on the total running time for the generic top-k weighted preference query.

Note that the top-k algorithm 4 is the only algorithm that is affected by the value of k. On the contrary, TA maintains a list of k elements. As k grows, the pruning benefit of TA and its variants decreases, and these algorithms become worst than sequential scan. Furthermore, since BSI computes and stores the scores for all the data points, it is suitable for interactive queries where users can increase the value of k to obtain immediate answers. It is also possible for the user to exclude data points from the result set as a bitArray. This bitArray can be used to initialized the E bitArray in the top-k Algorithm 4 to exclude those results.

Top-k Boolean Preference Queries. For top-k Boolean preference queries, Algorithm 2 does not need to be called, thus the cost of computing the top-k preferences is lower and it is given by:

$$O\left(|q|s\frac{n}{w}\right) \tag{7}$$

Top-k Baseline Preference Queries. In this case all the query weights are equal to one. The number of non-zero query weights equals to the number of data attributes: $|q| = m$. Again, in this case Algorithm 2 is not called. Thus the cost of top-k preference queries is:

$$O\left(ms\frac{n}{w}\right) \tag{8}$$

5   Experimental Evaluation

In this section we evaluate our approach for executing top-k queries over bit-sliced index structures. We first describe the experimental setup and the data sets used, which include a set of synthetic data sets as well as four real data sets. Then we show the benefits of using the BSI index structures in terms of compression ratios when compared to sorted lists. We then conduct a set of experiments, where we show the performance advantages of the BSI structures when compared to Sequential Top-k Algorithm (STA) [21], Threshold Algorithm (TA) [14], Best Position Algorithms (BPA, BPA2) [15], and Sequential Scan (SS). Because STA has shown to outperform TA and its optimizations, we compare BSI against STA in most of our experiments.

5.1   Experimental Setup

The synthetic data sets were generated using two different distributions: uniform and zipf. The cardinality of the generated data is 1,000 unless otherwise stated. The zipf distribution is representative for many real-world data sets, it is widely assumed to be ubiquitous for systems where objects grow in size or are fractured through competition [32]. These processes force the majority of objects to be small and very few to be large. Income distributions are one of the oldest exemplars first noted by Pareto [33] who considered their frequencies to be distributed as a power law. City sizes, firm sizes and word frequencies [32] have also been widely used to explore the relevance of such relations while more recently, inter-action phenomena associated with networks (hub traffic volumes, social contacts [34], [35]) also appear to mirror power-law like behavior. The zipf distribution generator uses a probability distribution of:

$$p(k, n, f) = \frac{1/k^f}{\sum_{i=1}^{n}(1/i^f)}$$

where n is the number of elements determined by cardinality, k is their rank, and the coefficient f creates an exponentially skewed distribution. We generated multiple data sets for f varying from 0 to 2. Further, we varied the number of rows, number of attributes as well as their cardinality to cover a large number of possible scenarios.

We also use four real data sets to support our results obtained with synthetic data:

- coil2000[1]. This data set used in the CoIL 2000 Challenge contains information on customers of an insurance company. Information about customers consists of 86 variables and includes product usage data and socio-demographic data derived from zip area codes. The data was supplied by the Dutch data mining company Sentient Machine Research and is based on a real world business problem. It contains over 9,000 descriptions of customers, including the information of whether or not they have a caravan insurance policy.
- internet[2] This data comes from a survey conducted by the Graphics and Visualization Unit at Georgia Tech October 10 to November 16, 1997. We use a subset of the data that provides general demographics of Internet users. It contains over 10,000 rows and 72 attributes with categorical and numeric values.
- kegg-metabolic[3] This is the KEGG Metabolic Relation Network (Directed) data set. It is a graph data, where Substrate and Product componds are considered as Edges while enzyme and genes are placed as nodes. There are 53,414 tuples in this data set, and 24 attributes, with real and integer values.
- poker-hand[4] In this data set each record is an example of a hand consisting of five playing cards drawn from a standard deck of 52. Each card is described using two attributes (suit and rank), for a total of 10 predictive attributes, plus one Class attribute that describes the "Poker Hand". The data set contains 1,025,010 instances and 11 attributes with categorical and numeric values.

For the experiments we generated three types of queries:

- Boolean queries: Every attribute of the query has a Boolean value (0 or 1), meaning that from the user perspective an attribute can be relevant or non-relevant.
- Baseline queries: Here every attribute is equally important for the user and hence the query vector is an all-ones vector.
- Weighted queries: Every attribute of the query has a weight between 0 and 1. The query weights are applied to each attribute before applying the top-k algorithm.

All experiments were executed over a machine with a 64-bit Intel Core i7-2600 processor (8MB Cache, 3.20 GHz) and 8 GB of memory, running Windows

---

[1] http://archive.ics.uci.edu/ml/machine-learning-databases/tic-mld/
[2] http://www.cc.gatech.edu/gvu/user_surveys/survey-1997-10
[3] http://archive.ics.uci.edu/ml/machine-learning-databases/00220/
[4] http://archive.ics.uci.edu/ml/datasets/Poker+Hand

7 Enterprise. Our code was implemented in Java. During the measurements, the queries were executed six times, and the result for the first run was discarded to prevent Java's *just-in-time* compilation from skewing results. The times from the other five runs were averaged and reported.

Query preferences are one decimal queries and follow a uniform random distribution. Each query set had 1,000 queries (unless otherwise noted). The query times reported corresponds to the average query time per query.

5.2 Index Size

Minimizing the index size is very important. A smaller index size results in less disk accesses and less main memory requirements. Often the index space reduction translates into faster processing of the index.

Figure 3 shows the index size of the original/raw data (SS, STA), the sorted lists (TA), and BSI using three different number of slices per attribute. For this experiment we generated a data set with 10 million tuples and varying number of attributes.

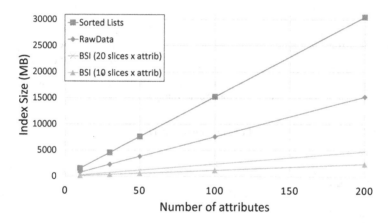

Fig. 3. Index size comparison for a data set with 10 million tuples and varying number of attributes. BSI is generated using 10 and 20 slices per attribute to represent 3 and 6 decimal numbers, respectively.

The storage requirement for the sorted lists is at least 2 times larger than the original data (assuming only the sorted lists are stored). The size further increases if the position list for each tuple is stored. In contrast, the total BSI size is smaller than the original data. If the normalized attribute values are represented using 3 decimal positions, then the number of slices used to represent each attribute is 10. In this case, for a 64-bit architecture, the BSI size is over 6 times smaller than the original data. When 6 decimals positions are used, 20 slices per attribute are generated but the index size is still one third the size of the original data.

5.3   BSI Performance Evaluation

In this section we evaluate our top-k query processing method using bit-sliced indices (BSI) by comparing it against the Sequential Top-k Algorithm (STA) [21], Threshold Algorithm (TA) [14], Best Position Algorithms (BPA, BPA2) [15], and Sequential Scan (SS). We designed a set of experiments that evaluate the usage of our approach for different types of data and queries.

Query Time vs. Data Dimensionality. Most of the existing indexing structures have been evaluated and are known to perform efficiently for a small number of dimensions [26]. A more scalable approach is Sequential Top-k Algorithm [21], it has been proven to outperform TA and its variants even for a higher number of dimensions.

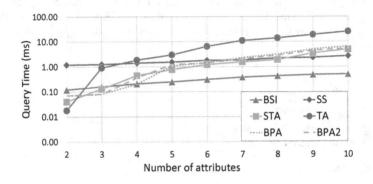

(a) Low-Dimensional Data

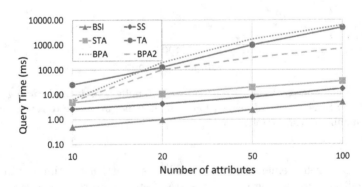

(b) High-Dimensional Data

Fig. 4. Query time for low and high-dimensional data. [Zipf-1, rows:100K, top-20].

Figures 4a and 4b show the query times for top-k preferences queries over a synthetically generated data set, where we vary the number of attributes. Figure

4a shows the query times for STA, BPA, BPA2, TA, SS, and BSI over a small number of attributes: varying from 2 to 10. For this experiment we extract top-20 candidates, and the data set has a Zipf-1 distribution.

STA performs slightly better than BSI when the number of attributes is 3 or smaller. Then, as the number of attributes grows STA query times grow at a faster rate than the query times for BSI. Note that BSI is always faster than sequential scan. TA, BPA and BPA2 are less scalable and their query times are even slower than STA, with TA having a response time of 25 ms for 10 dimensions (aprox. 100 times slower than BSI!).

To go even further and demonstrate the scalability of our approach, we show in Figure 4b the query times for up to 100 attributes, for the same data set. As can be seen, as the dimensionality grows, STA, TA, and BPA are outperformed by SS, while BSI remains over 3 times faster than SS.

One of the main advantages for using BSI is that it executes sequential scan over bit-slices when performing Algorithms 2, 3 and 4. Thus, it only accesses one data attribute at a time, and makes very efficient use of available memory and cache. Furthermore, due to this property, our approach has the potential to partition the BSI index and run Algorithms 2 and 3 in parallel.

Query Time vs. Number of Top Preferences. Figure 5 shows how the value of k (the number of preferences) impacts the query time for the preference query. The measurements were taken using a synthetically generated data-set with 10 million rows, 20 attributes and *zipf-1* data distribution. The queries used have a 1-decimal precision. Because BSI and sequential scan compute the score for all tuples regardless of the value of k, the time for computing top-k preferences is not significantly affected when increasing k. However, the query time for STA increases with the increase in the value of k as it takes longer to meet the stopping the criteria.

Query Time vs. Number of Tuples. To show the scalability of our approach, we present in Figure 6 the query times for a synthetically generated data set with a Zipf-1 distribution. We vary the number of rows for this data set from 10 thousand to 10 million rows. The number of attributes is 20, and the data has been normalized with 3 decimal precision. For this experiment we used a set of 100 weighted, 1-decimal, queries to extract top-20 candidates. In the results we present the average time taken to process a single query.

As can be seen in Figure 6, not only BSI is faster than both STA and SS, but also the query time per row remains constant for BSI (aprox. 7 ns/1K rows). This is not the case for STA, where the query time per row increases by adding more rows (45 ns/1K rows for a 10K rows data set , and 115 ns/1K rows for a 10M rows data set). Generally, as the number of rows grows, STA takes longer to reach its stopping criteria and becomes slower.

Query Time vs. Data Skewness. To show the effect of data distribution over BSI query processing, we vary the distribution of a data set from a uniform

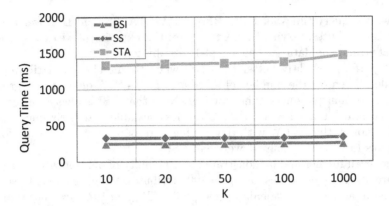

Fig. 5. Query times comparison when varying Top-k preferences. [Zipf-1, 20-attributes, 10M-rows].

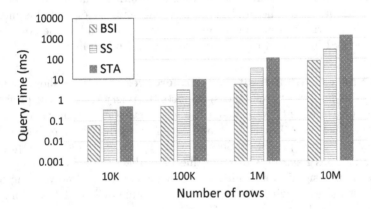

Fig. 6. Query times comparison when varying the number of rows. [Zipf-1, 20-attributes top-20].

(zipf-0) distribution to a Zipfian distribution with the skew factor $f = 2$ (zipf-2). The data set used in Figure 7 contains 100 thousand rows and 20 attributes with 10 bit-slices per attribute. The skewness of the data is towards the smaller values. For this experiment we use the same set of queries as in the previous one, for finding top-20 candidates.

As seen in Figure 7, BSI and Sequential Scan (SS) have constant processing times per query, irrespective of the data distribution, with BSI being more than twice as fast as SS. On the other hand, STA is highly sensitive to the data distribution and the query times are comparable to BSI only for highly skewed data. However, most real data sets have a skew factor lower than two. Zipf's Law states that the frequency of terms in a corpus conforms to a power law distribution where f is around 1 [36]. Also the same tendency has been observed in network graph data [34].

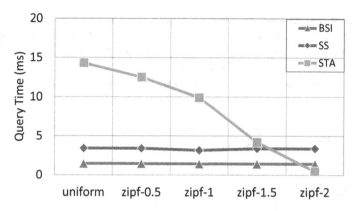

Fig. 7. Query time relative to data distribution. [100K-rows, 20-attributes, top-20].

Query Time vs. Attribute Cardinality. The cardinality of the data represents an important aspect for the BSI index. The higher the cardinality, the more slices need to be created per attribute. Figure 8 shows the query times for a uniformly distributed, synthetically generated data set that varies the attribute cardinality from 10 to 1,000,000. The data set has 10 million rows and 5 attributes. We used a set of 100 weighted queries to extract top-20 candidates. We use 1-decimal queries.

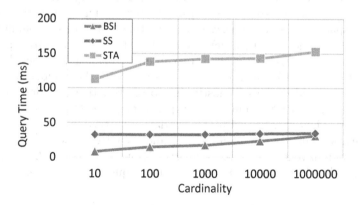

Fig. 8. Query time for varying data cardinality. [Uniform, 10M-rows, 5-attributes, top-20].

Figure 8 shows that the query times for both, STA and BSI, are dependent on the cardinality of the data. Both perform similarly with the increase in cardinality. However, for normalized data that is up to 6 decimals, BSI is still faster than SS, while STA is more than 4 times slower than SS. For lower data cardinalities, BSI multiplication algorithm 2 performs fewer iterations and thus translate to faster query times.

Query Time vs. Query Sparsity. Figure 9 compares the query time performance for SS, STA and BSI using sparse and non-sparse weighted queries. We vary the non-zero query attributes from 1% to 100%. For this measurement we use a synthetically generated data set with uniform distribution. It contains 100,000 rows and 1,000 attributes with 3 decimal precision.

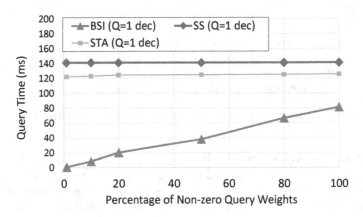

Fig. 9. Query time for sparse queries. [Uniform, 100K-rows, 1K-attributes, top-20].

Both, Sequential Scan (SS) and Sequential Top-k Algorithm (STA) scan sequentially the table for extracting top-k preferences. This means that regardless of the sparseness of the query, the query time will still be dominated by the scan of the table. On the other hand, TA and BSI are able to exploit the query sparseness, however TA is about two orders of magnitude slower than BSI. BSI query times increase only linearly-proportional with the number of non-zero query attributes added. This is mostly due to the fact that BSI does not need to load the entire index in memory when performing weight multiplication and finding top-k. Thus it works with smaller data structures and can also better exploit the availability of the main memory and the CPU cache. BSI Algorithms 2 and 3 are only invoked for non-zero query attributes. Hence, high-dimensional data is treated as low dimensional data when the queries are sparse.

Query Time vs. Query Slices. The results in Figure 10 are obtained using the same data set as in Figure 9. Here we look closer how BSI performs for different types of preference queries. More precisely, we vary the number of non-zero query attributes for Boolean queries, and for weighted queries with 1,2 and 3 decimal weights. The Boolean queries become all-ones queries(Baseline preference) for 100% non-zero boolean query weights. As the figure shows, by increasing the number of decimals for queries, the BSI Algorithm 2 will run slower, and this is reflected in the query times. However, for the queries with 50% or less non-zero attribute weights (500 attributes queried), BSI is still at least 50% faster than SS and about 25% faster than STA for 3-decimal queries.

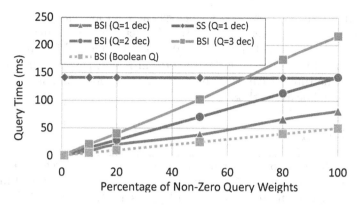

Fig. 10. Query time for different types of queries. [Uniform, 100K-rows, 1K-attributes, top-20].

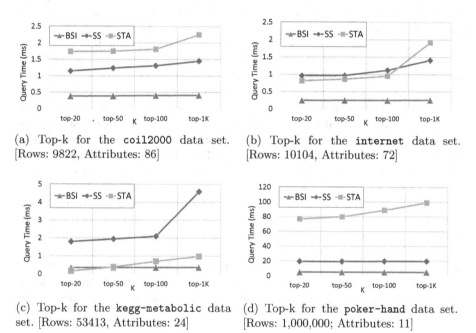

(a) Top-k for the `coil2000` data set. [Rows: 9822, Attributes: 86]

(b) Top-k for the `internet` data set. [Rows: 10104, Attributes: 72]

(c) Top-k for the `kegg-metabolic` data set. [Rows: 53413, Attributes: 24]

(d) Top-k for the `poker-hand` data set. [Rows: 1,000,000; Attributes: 11]

Fig. 11. Top-k weighted query on real data (K= 20 - 1,000)

Results for Real Datasets. Figure 11 shows the query times for the four real data sets described in section 5.1 as the number of query results increases (increasing k). For querying these data sets we use a set of $1,000$ weighted queries, uniformly generated, with 1-decimal query weights.

As the figure shows, increasing k when extracting top-k candidates, does not impact significantly the query time for BSI. In fact, the query time increases by only 0.01 - 0.03 ms when changing $k = 10$ to $k = 1000$ for all four data sets.

The reason is that the top K BSI algorithm is only invoked once after the scores for all the tuples have been computed. As expected, SS and STA performance increases linearly with k.

6   Conclusion

We have introduced a novel algorithm to perform top-k and preferences queries using bit-sliced indices for high-dimensional data. With the proposed indexing technique, there is no need to maintain a list of sorted attributes and it is easy to combine top-k queries with other types of queries for which bitmap indices have been traditionally recognized for, such as point and range queries. BSI uses only the number of bit-slices required by data value-range, thus in general, the index size is always smaller than, the size of the raw data.

In addition since attributes are indexed independently, our techniques can take advantage of the columnar storage and have the potential to be executed in parallel. We introduce several algorithms for processing top-k, and top-k weighted queries while exploiting the fast bit-wise operations enabled by the BSI index. This approach is robust and scalable for high dimensional data. In our experimental evaluation we show that by increasing the dimensionality of the data, BSI query times increase only linearly-proportional to the number of attributes added. Moreover, the distribution of the data does not affect the query performance for BSI, while TA and other threshold algorithms using sorted lists are very sensitive to data distribution.

Since the index is not sorted, updates to the index are more efficient than when keeping a sorted list. Also, since all the bits are sliced it is possible to further control the precision of the result and the query time performance by shedding the least significant bits. Further investigations are required for this matter, and will be a future research direction. Another future work direction is the exploration of bitmap compression for top-k queries. This can further reduce the space requirement of the BSI indices and still allow for fast bit-wise logical operations.

Acknowledgments. We would like to thank reviewers for their insightful comments on the paper, as these comments led us to an improvement of the work.

References

1. Ilyas, I.F., Beskales, G., Soliman, M.A.: A survey of top-k query processing techniques in relational database systems. ACM Comput. Surv. 40(4), 11:1–11:58 (2008). doi:10.1145/1391729.1391730. http://doi.acm.org/10.1145/1391729.1391730
2. Pagani, M.: Encyclopedia of Multimedia Technology and Networking, 2nd edn., Information Science Reference - Imprint of: IGI Publishing, Hershey (2008)

3. Böhm, C., Berchtold, S., Keim, D.A.: Searching in high-dimensional spaces: Index structures for improving the performance of multimedia databases. ACM Comput. Surv. 33(3), 322–373 (2001). doi:10:1145=502807:502809. http://doi.acm.org/10.1145/502807.502809

4. Daoudi, I., Ouatik, S.E., Kharraz, A.E., Idrissi, K., Aboutajdine, D.: Vector approximation based indexing for high-dimensional multimedia databases (2008)

5. Chaudhuri, S., Gravano, L., Marian, A.: Optimizing top-k selection queries over multimedia repositories. IEEE Trans. on Knowl. and Data Eng. 16(8), 992–1009 (2004). doi:10.1109/TKDE.2004.30. http://dx.doi.org/10.1109/TKDE.2004.30

6. Fagin, R.: Combining fuzzy information from multiple systems (extended abstract). In: Proceedings of the Fifteenth ACM SIGACT-SIGMOD-SIGART Symposium on Principles of Database Systems, PODS 1996, pp. 216–226. ACM, New York (1996). doi:10.1145/237661.237715. http://doi.acm.org/10.1145/237661.237715

7. Long, X., Suel, T.: Optimized query execution in large search engines with global page ordering. In: Proceedings of the 29th International Conference on Very Large Data Bases, VLDB 2003, VLDB Endowment, vol. 29, pp. 129–140 (2003)

8. Persin, M., Zobel, J., Sacks-davis, R.: Filtered document retrieval with frequency-sorted indexes. Journal of the American Society for Information Science 47, 749–764 (1996)

9. Cao, P., Wang, Z.: Efficient top-k query calculation in distributed networks. In: Proceedings of the Twenty-third Annual ACM Symposium on Principles of Distributed Computing, PODC 2004, pp. 206–215. ACM, New York (2004). doi:10.1145/1011767.1011798. http://doi.acm.org/10.1145/1011767.1011798

10. Wu, M., Xu, J., Tang, X., Lee, W.-C.: Top-k monitoring in wireless sensor networks. IEEE Trans. on Knowl. and Data Eng. 19(7), 962–976 (2007). doi:10.1109/TKDE.2007.1038

11. Balke, W.-T., Nejdl, W., Siberski, W., Thaden, U.: Progressive distributed top-k retrieval in peer-to-peer networks. In: Proceedings of the 21st International Conference on Data Engineering, ICDE 2005, pp. 174–185. IEEE Computer Society, Washington, DC (2005). doi:10.1109/ICDE.2005.115. http://dx.doi.org/10.1109/ICDE.2005.115

12. Metwally, A., Agrawal, D., Abbadi, A.E.: An integrated efficient solutionfor computing frequent and top-k elements in data streams. ACM Trans. Database Syst. 31(3), 1095–1133 (2006). doi:10.1145/1166074.1166084. http://doi.acm.org/10.1145/1166074.1166084

13. Marian, A., Bruno, N., Gravano, L.: Evaluating top-k queries over web-accessible databases. ACM Trans. Database Syst. 29(2), 319–362 (2004). doi:10.1145/1005566.1005569. http://doi.acm.org/10.1145/1005566.1005569

14. Fagin, R., Lotem, A., Naor, M.: Optimal aggregation algorithms for middleware. In: PODS, pp. 102–113 (2001)

15. Akbarinia, R., Pacitti, E., Valduriez, P.: Best position algorithms for top-k queries. In: Proceedings of the 33rd International Conference on Very Large Data Bases, VLDB 2007, VLDB Endowment, pp. 495–506. http://dl.acm.org/citation.cfm?id=1325851.1325909

16. Yu, A., Agarwal, P.K., Yang, J.: Topk preferences in high dimensions (2014)

17. Gurský, P., Vojtáš, P.: Speeding up the nra algorithm. In: Greco, S., Lukasiewicz, T. (eds.) SUM 2008. LNCS (LNAI), vol. 5291, pp. 243–255. Springer, Heidelberg (2008). http://dx.doi.org/10.1007/978-3-540-87993-0_20

18. Mamoulis, N., Cheng, K.H., Yiu, M.L., Cheung, D.W.: Efficient aggregation of ranked inputs. In: ICDE. IEEE Computer Society, p. 72 (2006)

19. Natsev, A., Chang, Y.C., Smith, J.R., Li, C.-S., Vitter, J.S.: Supporting incremental join queries on ranked inputs. In: VLDB, pp. 281–290 (2001)

20. Güntzer, U., Balke, W.-T., Kießling, W.: Optimizing multi-feature queries for image databases, pp. 419–428 (2000)

21. Jin, W., Patel, J.M.: Efficient and generic evaluation of ranked queries. In: Proceedings of the 2011 ACM SIGMOD International Conference on Management of Data, SIGMOD 2011, pp. 601–612. ACM, New York (2011). doi:10.1145/1989323.1989386. http://doi.acm.org/10.1145/1989323.1989386

22. O'Neil, P., Quass, D.: Improved query performance with variant indexes. In: Proceedings of the 1997 ACM SIGMOD International Conference on Management of Data, pp. 38–49. ACM Press (1997). http://doi.acm.org/10.1145/253260.253268

23. Rinfret, D., O'Neil, P., O'Neil, E.: Bit-sliced index arithmetic. SIGMOD Rec. 30(2), 47–57 (2001). doi:http://doi.acm.org/10.1145/376284.375669

24. Wu, M.-C., Buchmann, A.P.: Encoded bitmap indexing for data warehouses. In: ICDE 1998: Proceedings of the Fourteenth International Conference on Data Engineering, pp. 220–230. IEEE Computer Society Washington, DC (1998)

25. Fagin, R., Kumar, R., Sivakumar, D.: Comparing top k lists. In: Proceedings of the Fourteenth Annual ACM-SIAM Symposium on Discrete Algorithms, SODA 2003, pp. 28–36. Society for Industrial and Applied Mathematics. Philadelphia (2003). http://dl.acm.org/citation.cfm?id=644108.644113

26. Pang, H., Ding, X., Zheng, B.: Efficient processing of exact top-k queries over disk-resident sorted lists. The VLDB Journal 19(3), 437–456 (2010). doi:10:1007=s00778–009–0174–x. http://dx.doi.org/10.1007/s00778-009-0174-x

27. Bast, H., Majumdar, D., Schenkel, R., Theobald, M., Weikum, G.: Io-top-k: Index-access optimized top-k query processing, In: Proceedings of the 32nd International Conference on Very Large Data Bases, VLDB 2006, VLDB Endowment, pp. 475–486 (2006). http://dl.acm.org/citation.cfm?id=1182635.1164169

28. Gurský, P., Vojtáš, P.: On Top-k search with no random access using small memory. In: Atzeni, P., Caplinskas, A., Jaakkola, H. (eds.) ADBIS 2008. LNCS, vol. 5207, pp. 97–111. Springer, Heidelberg (2008). http://dx.doi.org/10.1007/978-3-540-85713-6_8

29. Chuan Chang, K.C., won Hwang, S.: Minimal probing: Supporting expensive predicates for top-k queries. In: SIGMOD, pp. 346–357 (2002)

30. Das, G., Gunopulos, D., Koudas, N., Tsirogiannis, D.: Answering top-k queries using views. In: Proceedings of the 32nd International Conference on Very Large Data Bases, VLDB 2006, VLDB Endowment, pp. 451–462 (2006). http://dl.acm.org/citation.cfm?id=1182635.1164167

31. Cong, G., Jensen, C.S., Wu, D.: Efficient retrieval of the top-k most relevant spatial web objects

32. Clauset, A., Shalizi, C.R., Newman, M.E.J.: Power-law distributions in empirical data (2009). doi:10.1137/ 070710111. http://dx.doi.org/10.1137/070710111

33. Pareto, V.: Manual of political economy (1906)

34. lászló Barabáasi, A., Albert, R.: Emergence of scaling in random networks, Science

35. Barabasi, A.-L.: The origin of bursts and heavy tails in human dynamics. Nature 435, 207 (2005). http://www.citebase.org/abstract?id=oai:arXiv.org:cond-mat/0505371

36. Zipf, G.: Human behaviour and the principle of least-effort. Addison-Wesley, Cambridge (1949). http://publication.wilsonwong.me/load.php?id=233281783

SeeVa: A Model Based Framework for Semantic Web Service Discovery

Roberto De Virgilio[1]([⊠]) and Devis Bianchini[2]

[1] Dipartimento di Ingegneria, Università Roma Tre, Rome, Italy
dvr@dia.uniroma3.it
[2] Dipartimento di Ingegneria dell'Informazione, Università degli Studi di Brescia,
Brescia, Italy
bianchin@ing.unibs.it

Abstract. Semantic Web service (SWS) discovery has gained more and more attention, leading to a great number of service matchmaking approaches. Existing approaches are based on SWS descriptions expressed according to a single specification (e.g., OWL-S, WSMO and SAWSDL). In this paper we propose a service matchmaking algorithm based on a SWS meta-model that abstracts the features of all the most common SWS specifications. The algorithm performs SWS comparison by increasingly relaxing matchmaking constraints, in order to maximize effectiveness of the discovery procedure, in terms of precision and recall. Moreover, to speed up algorithm performances, we provide SeeVa, an efficient representation of the SWS meta-model on which the algorithm is based. SeeVa is a storage system that includes a Datalog engine to enable language-independent reasoning capabilities. We evaluate the algorithm on public datasets containing SWS descriptions expressed using different specifications. Experiments demonstrate how the proposed approach outperforms main existing service matchmaking solutions both in terms of precision and recall and in terms of response time, thanks to the storage system and the Datalog engine.

1 Introduction

Web services promote easy access to remote content and application functionality, independent of the provider's platform and the service implementation. Available Web services are made accessible by providers to (Web service) requesters through the specification of their interface (i.e., the operations, input/output and fault messages), their bindings (i.e., the networking details about how the messages must be expressed to interact with the service) and the endpoints where a service is available, using the Web Service Description Language (WSDL). Web service interface description is exploited by requesters to find Web services, aggregate them as components to form composite services, and invoke them according to the details provided within WSDL bindings.

Automatic Web service discovery becomes a crucial step to pursue the composition and invocation of Web services [34]. It enables the automatic retrieval

© Springer-Verlag Berlin Heidelberg 2014
A. Hameurlain et al. (Eds.): TLDKS XIV, LNCS 8800, pp. 51–82, 2014.
DOI: 10.1007/978-3-662-45714-6_3

of Web services, hereafter denoted with *Web service advertisements*, that match a given *Web service request* (i.e., the description of a desired Web service as formulated by the requester). Since requesters have access to the Web service interface for performing search, service matchmaking is mainly based on comparison between interface elements. Performing keyword-based comparison to find Web services is hampered by the inner ambiguity of keywords, such as polisemy (i.e., the same term refers to different concepts) and synonymy (i.e., the same concept is pointed out using different terms). In Web resource discovery in general, polisemy increases the number of false positives (i.e., non-relevant services included among search results), while synonymy increases the number of false negatives (i.e., relevant services not included among search results). False positives negatively affect the search precision, while false negatives worsen the search recall. To address these issues, Semantic Web technologies [32] have been widely applied. Semantic Web technologies conceptualize semantics of Web service descriptions with concepts explicitly defined in (domain) ontologies. When automatic service discovery is based on Semantic Web technologies, we denote is as *Semantic Web Service (SWS) discovery*. It has been widely addressed, but it is still a challenging research topic [27]. Several semantic Web service discovery approaches have been proposed in literature [2,21,24], based on different ways to conceptualise service semantics [3,14,26] (see Section 2). Different approaches also apply distinct techniques for determining the degree of match between the request and the advertisement. Logic-based approaches [15,17,24] use reasoning techniques based on Description Logics (that mainly rely on the tableau algorithm [30]) to distinguish among exact matches and mismatches. Similarity-based approaches [13,28] apply techniques from Information Retrieval (IR) and concept distance over ontologies/thesauri. Hybrid approaches [2,19,21, 22] use reasoning techniques of logic-based ones, but also tolerate partial/approximate match through the application of similarity-based techniques. Hybrid solutions add flexibility (thus increasing recall of search results) to the precision of the logic-based techniques. Our aim in this paper is to propose a hybrid service matchmaking algorithm based on a semantic Web service meta-model that abstracts the features of all the most common SWS specifications. The meta-model is in turn built upon a meta-meta-model defined in [9] to represent the primitives of Semantic Web ontology languages (e.g., the Ontology Web Language, OWL). This enables to provide an efficient representation of the model within SeeVa[1], a storage system based on a Datalog engine that includes language-independent reasoning capabilities on a relational implementation of the constructs used to represent both the semantic Web service descriptions and the ontology used to conceptualise their semantics. We evaluate the algorithm on public datasets containing SWS descriptions expressed using different specifications.

In [8], we provided a first solution to the problem by relying on the representation of the WSDL document using the constructs of the meta-meta-model. We also proposed a service matchmaking approach that is purely based on logic

---

[1] SeeVa is the sanskrit transliteration of the english word *service*.

reasoning on the meta-meta-model. Here we extended that work as follows: (i) we define the *semantic Web service meta-model*; (ii) we describe the service matchmaking algorithm; (iii) we formulate the algorithm using the constructs of the meta-meta-model and we translate it as a set of Datalog programs and rules, in order to efficiently perform service matchmaking.

The paper is organized as follows. Section 2 provides some preliminary definitions and examples useful to understand the next sections of the paper. Section 3 describes the semantic Web service meta-model and how it is represented within the SeeVa storage system. The hybrid service matchmaking algorithm is detailed in Section 4. Section 5 provides implementation issues and the evaluation of the algorithm, by means of: (i) a comparison with related work to clarify the cutting-edge features of our approach compared to existing solutions, (ii) the description of performed experiments. Finally, future work and concluding remarks are discussed in Section 6.

2   Background and Examples

Almost all the semantic Web service discovery approaches rely on semantic Web service specifications like OWL-S (Ontology Web Language for Services) [26], WSMO (Web Service Modeling Ontology) [3] and SAWSDL (Semantic Annotations for WSDL) [14]. Although some solutions are based on ad-hoc semantic Web service representations [2,17], they also extract such a representation from semantic Web service descriptions expressed in OWL-S, WSMO or SAWSDL. Therefore, we can consider the features of these three specifications for designing our service meta-model without loss of completeness. The most widely adopted solutions for semantic Web service description provide an upper level ontology (OWL-S) and a conceptual model (WSMO) to represent Web services and their semantic annotations or start from the WSDL specification and extend it with semantic annotations with concepts taken from a domain-specific ontology (SAWSDL).

Semantic Web Service Specifications. Specifically, *semantic Web services* are Web service descriptions enriched with: (i) the specification of semantic meaning of Web service elements (e.g., operations, input/output and fault messages), pursued through the semantic annotation of such elements with concepts taken from *domain ontologies* (hereafter, *ontologies*); (ii) semantic constraints as logical expressions to describe the state of the domain before the execution of the Web service (pre-conditions) and after its execution (post-conditions). We envision a registry where semantic Web service advertisements are made available to requesters. For introducing the examples of this section we defined an ontology, TravelOnt, whose TBox is partially formalized in Fig. 1 using Description Logics. It is up to the community of Web service providers and requesters, with the help of an expert in ontology representation formalisms, to create such an ontology depending on their needs.

Accomodation ⊑ ∃locatedIn.Destination ⊓ ∃reservedThrough.Booking ⊓
    ∃hasActivity.Activity ⊓ ∃reservedBy.Reservation,
Bed&Breakfast ⊑ Accomodation, Hotel ⊑ Accomodation,
LuxuryHotel ⊑ Hotel, Lodge ⊑ Accomodation,
UrbanArea ⊑ Destination, Town ⊑ UrbanArea,
City ⊑ UrbanArea, Capital ⊑ City,
RuralArea ⊑ Destination, NationalPark ⊑ RuralArea, Beach ⊑ Destination,
B&BBooking ⊑ Booking, LodgeBooking ⊑ Booking,
Relax ⊑ Activity, Spa ⊑ Relax, Sauna ⊑ Relax, FinnishSauna ⊑ Sauna,
Sport ⊑ Activity, Surfing ⊑ Sport, Hiking ⊑ Sport,
Reservation ⊑ ∃totalAmount.Price ⊓ ∃payment.PaymentMethod ⊓
    ∃includes.Service,
CreditCard ⊑ PaymentMethod, Breakfast ⊑ Service

Fig. 1. A portion of the `TravelOnt.owl` ontology (expressed using Description Logics), used for the semantic annotation of Web services in the running example

OWL-S (Ontology Web Language for Services). The OWL-S proposal defines an upper level OWL service ontology, where a semantic Web Service is described by three properties: it *presents* a *ServiceProfile*, describing what the service does, it is *describedBy* a *ServiceModel*, describing how the service works, and it *supports* a *ServiceGrounding* for its invocation. The *ServiceProfile* contains information about the functionality the service provides (*functional description*) and the service category within any classification systems (e.g., UNSPSC[2]), which are used for Web service discovery purposes. The functional description characterizes the service functionality in terms of the information transformation (represented by Inputs and Outputs, IO) and the state change produced by the execution of the service (represented by Preconditions and Effects, PE). The OWL-S *ServiceProfile* does not contain a schema to describe IOPE: such a schema is provided within the *ServiceModel* and represents the ontological concepts that are associated to the IO and expressions associated to PE.

Example. The semantic Web service advertisement shown in Fig. 2 is expressed using the OWL-S specification. The service has two inputs (`City`,`CreditCard`) and three outputs (`LuxuryHotel`, `Sauna` and `Spa`), associated to concepts with the same name defined in the travel ontology. OWL-S ontology also provides means to organize service profiles in order to position a service within a classification or a broader array of services to further facilitate their discovery. The most natural way to do this is to build a profile hierarchy, represented as an OWL ontology (more details can be found in http://www.daml.org/services/owl-s/1.0/ProfileHierarchy.html). For instance, in Fig. 3 the `LuxuryHotel_profile` is defined as a subclass of `Booking` services.

---

[2] United Nations Standard Products and Services Code® (UNSPSC®):
http://www.unspsc.org

```
<rdf:RDF <!ENTITY travel "http://localhost:8080/TravelOnt.owl">
 <Service rdf:ID="LuxuryHotel_Booking_service">
  <presents rdf:resource="#LuxuryHotel_profile"/>
  <describedBy rdf:resource="#LuxuryHotel_process"/>
  <supports rdf:resource="#LuxuryHotel_grounding"/>
 </Service>
 <profile:Profile rdf:ID="LuxuryHotel_profile">
  <profile:hasInput rdf:resource="#CITY"/>
  <profile:hasInput rdf:resource="#CREDIT_CARD"/>
  <profile:hasOutput rdf:resource="#LUXURY_HOTEL"/>
  <profile:hasOutput rdf:resource="#SAUNA"/>
  <profile:hasOutput rdf:resource="#SPA"/>
 </profile:Profile>
 <process:Input rdf:ID="CITY">
  <process:parameterType rdf:datatype="anyURI">
     &travel;#City
  </process:parameterType>
 </process:Input>
 <process:Input rdf:ID="#CREDIT_CARD">
  <process:parameterType rdf:datatype="anyURI">
     &travel;#CreditCard
  </process:parameterType>
 </process:Input>
 <process:Output rdf:ID="#LUXURY_HOTEL">
  <process:parameterType rdf:datatype="anyURI">
     &travel;#LuxuryHotel
  </process:parameterType>
 </process:Output>
 <process:Output rdf:ID="SAUNA">
  <process:parameterType rdf:datatype="anyURI">
     &travel;#Sauna
  </process:parameterType>
 </process:Output>
 <process:Output rdf:ID="SPA">
  <process:parameterType rdf:datatype="anyURI">
     &travel;#Spa
  </process:parameterType>
 </process:Output>
            . . .
```

Fig. 2. An example of Semantic Web service description expressed using OWL-S

```
<rdf:RDF
 <!ENTITY profile "http://www.daml.org/services/
     owl-s/1.0/Profile.owl">
 <!ENTITY travel "http://localhost:8080/TravelOnt.owl">
            . . .
 <owl:Class rdf:ID="&travel;#Booking">
   <rdfs:subClassOf rdf:resource="&profile;#Profile"/>
 </owl:Class>
 <owl:Class rdf:ID="LuxuryHotel_profile">
   <rdfs:subClassOf rdf:resource="&travel;#Booking"/>
 </owl:Class>
            . . .
</rdf:RDF>
```

Fig. 3. An example of Web service profile hierarchy using OWL-S

WSMO (Web Service Modeling Ontology). In the WSMO proposal a semantic Web Service is described by four top-level elements: (i) *ontologies*, to provide the terminology used by other WSMO elements to describe the relevant aspects of the domain of interest; (ii) *Web services*, to describe the computational entity that provides some values in the domain (they correspond to advertisements); (iii) *goals*, to represent the Web service request; (iv) *mediators*, to deal with interoperability problems between different WSMO elements. Goals and Web services are described by *capabilities*, used for discovery purposes. Capabilities define Web services by means of: pre-conditions and post-conditions (to specify the information space of the Web service before and after its execution), assumptions and effects (to describe the state of the domain before and after the execution of the Web service). Pre-conditions and post-conditions are simple axioms to denote that IO are instances of given classes within the ontology and correspond to IO semantic annotation. Assumptions and effects correspond to OWL-S PE.

Example. Let us consider the semantic Web service advertisement shown in Fig. 4. In the example, the travel ontology is imported and assigned to the `travel` namespace prefix. The Web service presents two inputs, annotated with the `City` and `CreditCard` concepts (using pre-condition axioms) and two outputs, annotated with the `Accomodation` and `FinnishSauna` concepts (using post-condition axioms). Axioms are logical expressions represented in WSML (Web Service Modeling Language) [20]. The notation `?x` is used to denote variables. The counterpart of OWL-S profile hierarchy in WSMO is achieved through mediators, that can be used to link Web services, thus enabling to define new Web services by refining existing ones. For instance, a new Web service can be defined and a mediator can be used to state that it refines the Web service shown in Fig. 4 by adding more constraints in terms of WSML axioms.

SAWSDL (Semantic Annotations for WSDL). According to the SAWSDL proposal, semantic annotation is performed through a `modelReference` attribute

---

namespace travel: <<http://localhost:8080/TravelOnt.owl>>
webservice <<http://...Booking.wsml>>
  importOntologies <<http://...TravelOntology.owl>>
  capability
    sharedVariables ?City, ?CreditCard, ?Accomodation, ?FinnishSauna
    precondition definedBy
      ?City **memberOf** travel:City
    precondition definedBy
      ?CreditCard **memberOf** travel:CreditCard
    postcondition definedBy
      ?Accomodation **memberOf** travel:Accomodation
    postcondition definedBy
      ?FinnishSauna **memberOf** travel:FinnishSauna

Fig. 4. An example of semantic Web service description expressed using WSMO

that associates a concept within an ontology to: (i) the `wsdl:interface` tag to annotate the whole Web service (e.g., through a category taken from a service categorization); (ii) the `wsdl:operation` tag to annotate an operation; (iii) any `element` tag within the `types` part of the specification, to annotate Web service IO. Semantic annotations obtained through the `modelReference` attribute are used for semantic Web service discovery. PE can be added to the SAWSDL specification by referencing, through the `modelReference` attribute, external documents where conditions are defined, without any commitment on the adopted logical formalism.

Example. For instance, in the semantic Web service description shown in Fig. 5, expressed using the SAWSDL specification, the operation is annotated using the `LodgeBooking` concept, presents two inputs, annotated with the concepts `City` and `Room`, and only one output, annotated with the concept `Lodge`.

Web Service Matchmaking. *Web service matchmaking* is defined as a process that requires a repository of *Web service advertisements* and a *Web service request* as input, and returns all advertisements that may potentially match the requirements specified in the Web service request [17]. Most Web service matchmakers use the IOPE model (Input-Output-Precondition-Effect) to represent Web service advertisements and requests, since they rely on the OWL-S ontology, the WSMO conceptual model or the SAWSDL specification. There can be different kinds of match between a service advertisement S and a request R. In literature, the kinds of match are formalized starting from the definition of *plug-in* match defined in [36]. According to this definition, a Web service advertisement plugs into the service request if, for each required output, the Web service advertisement provides an equivalent or more specific output and, for each input in the Web service advertisement, there is an equivalent or more specific input in the request R. This definition ensures that the inputs specified within the request R are enough to execute the advertisement S. Starting from the definition of plug-in match, other kinds of match are defined. In Section 4 we will describe the kinds of match we considered in our service matchmaking algorithm.

3   Semantic Web Service Model

The model we adopt to represent both Web service advertisements and requests aims at summarizing the elements from OWL-S, WSMO and SAWSDL that are commonly used for Web service discovery purposes.

Web Service Advertisements. Formally, we define a *Web service advertisement* S as a set composed of the following elements op_S:

$$op_S = \langle C_S^{op}, IN_S^{op}, OUT_S^{op} \rangle \tag{1}$$

```
<wsdl:description
  xmlns:wsdl="http://www.w3.org/ns/wsdl"
  xmlns:xs="http://www.w3.org/2001/XMLSchema"
  xmlns:sawsdl="http://www.w3.org/ns/sawsdl"
  xmlns:travel="http://locahost:8080/TravelOnt.owl">

<wsdl:types>
 <xs:schema>
  <xs:element name="BookingRequest">
   <xs:complexType>
    <xs:sequence>
     <xs:element name="City"
     sawsdl:modelReference="travel#City"/>
     <xs:element name="Room"
     sawsdl:modelReference="travel#Room"/>
    </xs:sequence>
   </xs:complexType>
  </xs:element>
  <xs:element name="BookingResponse">
   <xs:complexType>
    <xs:sequence>
     <xs:element name="Lodge"
     sawsdl:modelReference="travel#Lodge"/>
    </xs:sequence>
   </xs:complexType>
  </xs:element>
 </xs:schema>
</wsdl:types>

<wsdl:interface name="bookingInterface">
 <wsdl:operation name="booking"
 pattern="http://www.w3.org/ns/wsdl/in-out"
 sawsdl:modelReference="travel#LodgeBooking">
    <wsdl:input element="BookingRequest"/>
    <wsdl:output element="BookingResponse"/>
 </wsdl:operation>
</wsdl:interface>

<wsdl:service name ="reservationService"
interface="bookingInterface">
 ...
</wsdl:service>
</wsdl:description>
```

Fig. 5. An example of semantic Web service description expressed using SAWSDL

Each op_S element corresponds to: (i) an operation according to the SAWSDL specification; (ii) a service profile according to the OWL-S ontology; (iii) a capability according to the WSMO conceptual model. C_S^{op} represents a set of ontological concepts used to semantically characterize the whole op_S element. These concepts can be used to model: (a) the categories (common to all the op_S within the same advertisement S) if any Web service categorization schema exists within the repository of service advertisements (e.g., by means of profile hierarchies in the OWL-S ontology); (b) the concepts used to annotate interfaces according to the SAWSDL specification (common to all the op_S within the same advertisement S); (c) the concepts used to annotate operations according to the SAWSDL specification. The idea is that a semantic characterization of operations should be considered as a first, coarse-grained criteria to identify candidate services.

Categories can be used to semantically characterize a Web service and, within the same Web service, additional concepts, if any, can be considered to characterize each single operation. More in-depth semantic description of Web service advertisements is given on IO. IN_S^{op} (resp., OUT_S^{op}) represents a set of concepts used to annotate the (operation) inputs (resp., outputs). In Formula (1) we did not include the logical expressions that represent the pre- and post-conditions (according to the SAWSDL and OWL-S terminologies) or the assumptions and effects (according to the WSMO terminology), since, to the best of our knowledge, this holds for the hybrid matchmakers and there are no systems that implement a real integrated IOPE matching. Actually, the inclusion of (complex) logical expressions would require that semantic Web service providers are experienced in their formulation. This requirement could be considered for ad-hoc situations, that we will investigate as future work with a proper extension of our work.

Example. Let us consider the three examples of semantic Web service descriptions given in Section 2. The OWL-S description shown in Fig. 2 and Fig. 3 is modeled as a single-operation Web service $\mathcal{S}_1$:

$$op_{\mathcal{S}_1} = \langle \texttt{Booking},\{\texttt{City},\texttt{CreditCard}\},\{\texttt{LuxuryHotel},\texttt{Sauna},\texttt{Spa}\}\rangle$$

The WSMO description shown in Fig. 4 is modeled as a single-operation Web service $\mathcal{S}_2$:

$$op_{\mathcal{S}_2} = \langle \emptyset,\{\texttt{City},\texttt{CreditCard}\},\{\texttt{Accomodation},\texttt{FinnishSauna}\}\rangle$$

The SAWSDL description shown in Fig. 5 is modeled as a single-operation Web service $\mathcal{S}_3$:

$$op_{\mathcal{S}_3} = \langle\{\texttt{LodgeBooking}\},\{\texttt{City},\texttt{Room}\},\{\texttt{Lodge}\}\rangle$$

For clarity reasons, in these examples we referred to concepts using their names only. Actually, each concept is uniquely identified by its name within the domain ontology used for semantic annotation.

Web Service Request. A *Web service request* is defined as the set of requirements to find a given functionality. According to this vision, a request $\mathcal{R}$ is represented as a set $C_\mathcal{R}$ of concepts used for a coarse-grained semantic characterization of the functionality to search for and two sets of inputs ($IN_\mathcal{R}$) and outputs ($OUT_\mathcal{R}$) to better specify the required Web service:

$$\mathcal{R} = \langle C_\mathcal{R}, IN_\mathcal{R}, OUT_\mathcal{R}\rangle \tag{2}$$

where each concept is uniquely identified by its name within the domain ontology.

Example. To find a Web service that enables to book an hotel with sauna by specifying the city where the hotel is located and the credit card as payment method, $\mathcal{R}$ is formulated as follows, with reference to the `TravelOnt.owl` ontology:

$$\mathcal{R} = \langle \texttt{Booking}, \{\texttt{City}, \texttt{CreditCard}\}, \{\texttt{Hotel}, \texttt{Sauna}\} \rangle$$

During Web service matchmaking, $\mathcal{R}$ is matched against $op_\mathcal{S}$ of available Web service advertisements.

3.1   Model-Based Web Service Storage

We follow the approach proposed in [1] for the management of heterogeneous data models in a uniform way, where a meta-model, composed of a set of generic constructs, is defined to enable the representation of both meta-data and data according to the semantic Web service description shown in Formula (1) and primitives used in semantic Web specifications, such as the Resource Description Framework Schema (RDFS) and OWL. The adoption of the meta-model presents two main advantages: (i) it provides a framework in which different semantic Web specifications can be handled in a uniform way and (ii) it allows the definition of *language-independent* reasoning capabilities. In [9–11] we defined NYAYA, a system for the management of Semantic-Web data, which couples a general-purpose and extensible storage mechanism with efficient ontology reasoning and querying capabilities. In particular, NYAYA exploits a profitable relational implementation of the meta-model. This implementation enables to exploit indexing and partitioning techniques to guarantee good performances when the tables become very large. In this section we summarize the features of NYAYA. Our aim in this paper is to exploit the meta-model and the NYAYA implementation to store both the semantic Web service descriptions according to Formula (1) and the representation of ontologies used for Web service semantic annotation within the same storage system. This would enable efficient reasoning capabilities on datasets of semantic Web services expressed according to different languages. The innovative service matchmaking algorithm discussed in the next section partially relies on such a storage system, thus exploiting its reasoning capabilities.

The Meta-model. NYAYA proposes a simple meta-model $\mathcal{M}$, that can be represented as $\mathcal{M} = \{C_1, C_2, \ldots, C_n\}$, where C_i are constructs with the following structure:

$$C_i = (OID_i, attr_i^1, \ldots, attr_i^f, ref_i^1, \ldots, ref_i^m) \tag{3}$$

where OID_i is the object identifier, $attr_i^j$ is the j-th property of the construct C_i and ref_i^k is the k-th reference toward other constructs. In this section, we first present the constructs of the meta-model and then we detail how elements involved in the representation of both the Web Service interface (i.e. $\mathcal{S}$) and the semantic annotation (i.e. $\mathcal{A}$) are mapped into the contructs of $\mathcal{M}$. The constructs of $\mathcal{M}$ are logical predicates, each of which represents a primitive used in a Semantic Web language (e.g., RDFS and OWL) for describing the domain of interest. Each construct is associated with an object identifier, a name, a set of properties and a set of references to other constructs. The approach is basically independent of the actual storage model (e.g., relational, object-oriented, or XML-based). However, since the SeeVa storage system relies on a

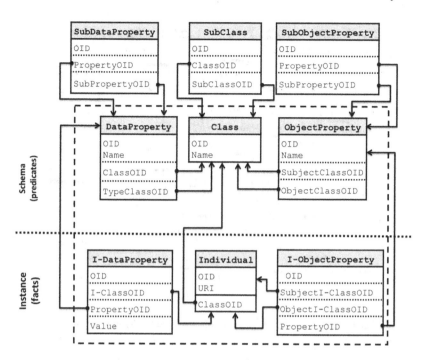

Fig. 6. A fragment of the meta-model used to represent both the ontology and the semantic Web service descriptions

relational implementation of the meta-model, in the following we assume that each construct corresponds to a relational table.

In Fig. 6 we show an UML-like diagram representing a fragment of $\mathcal{M}$ containing the main constructs of OWL-QL (OWL Query Language). Schema-level constructs are distinguished from instance-level constructs. The rectangles represent constructs and the arrows references between them. Notice how the core of $\mathcal{M}$ (enclosed in the dashed box) can serve to represent the facts of both OWL-QL and RDFS(DL) ontologies, that contain the set of RDFS constructs that can be expressed using Description Logics, and therefore are tractable. In particular the DATAPROPERTY table has a reference to the CLASS it belongs to and has the range data type (e.g., integer or string) as attribute. The OBJECTPROPERTY relation is used to represent RDF statements between individuals and has two references to the subject class and the object class involved in the statement. In addition, it has attributes that specify notable features of the property. The INDIVIDUAL relation is used to model resources that are instances of the class specified by its ClassOID property. I-DATAPROPERTY and I-OBJECTPROPERTY have been defined in a similar way. The full specification of the meta-model embeds several others constructs; exploiting the notion of meta-model, a strong point of the approach is the extendibility. If the meta-model is not recognized as expressive enough for user's purposes, new constructs (with references and

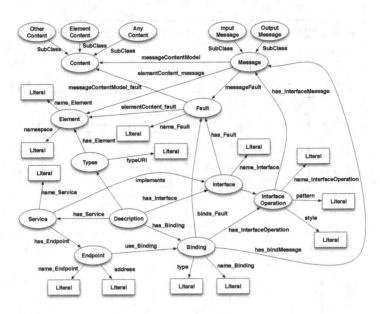

Fig. 7. Semantic representation of the WSDL Schema elements

properties) and/or new properties can be added. However for our purposes, in the following we will refer to the fragment shown in Fig. 6. The adoption of a meta-model able to describe all the language constructs of interest has two main motivations. On the one hand, it provides a framework in which Semantic Web languages can be handled in a uniform way, enabling to switch between the different languages and to define *language-independent* reasoning capabilities [9,10]. On the other hand the separation of concerns, employed into the approach, allows for modeling data and metadata at different layers (e.g., conceptual, logical and physical).

The Meta-model in Action: Representing Semantic Web Services. We characterize the set of basic constructs to describe a WSDL document, denoted with $\mathcal{M}_{WSDL}$, we select the fragment of the meta-model $\mathcal{M}$ illustrated in Fig. 6, denoted with $\mathcal{M}_{fragment}$, and we define a mapping between the two collections of constructs. This mapping represents a *semantic representation* of $\mathcal{S}$. Formally, we define $\mathcal{M}_{WSDL}$ at the conceptual layer through the following constructs: $C_{description}$, C_{types}, $C_{interface}$, $C_{binding}$ and $C_{service}$.

Such model represents the core set of constructs to describe a WSDL document. The main component is $C_{description}$, which presents the attribute *targetNamespace* to identify the WSDL components occurring in the document. $C_{description}$ is related to four top-level elements: C_{types} describes the format of messages used by the service as inputs and outputs; $C_{interface}$ describes the abstract the functionality of the service; $C_{binding}$ specifies the implementation details necessary to access the service, that is, a concrete message format and communication protocols of the interface for each operation; finally, $C_{service}$ describes where the service can be accessed. We enrich $\mathcal{M}_{WSDL}$ adding other constructs such as $C_{operation}$, $C_{message}$ and so on.

The semantic representation of a WSDL document at the conceptual layer results from the mapping between $\mathcal{M}_{WSDL}$ and $\mathcal{M}_{fragment}$. Fig. 7 shows the resulting semantic representation of the elements in the abstract part of the WSDL. We used a graph representation where nodes correspond to CLASS, red edges to OBJECTPROPERTY, green edges to DATAPROPERTY and purple edges to SUBCLASS relations between classes. The semantic representation of the WSDL Binding elements at schema level follows the same notation and has been omitted, since our matchmaking approach does not rely on binding elements, as for the other service matchmakers proposed in literature.

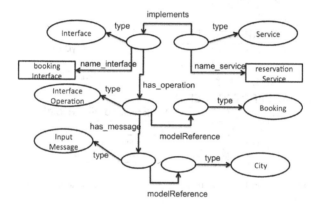

Fig. 8. Semantic representation of a sample service

Example. Let us consider the running example in Section 2. Fig. 8 shows the resulting semantic representation of the service S_1 at conceptual layer considering instance-level constructs, where nodes correspond to INDIVIDUAL and CLASS, while edges correspond to I-OBJECTPROPERTY and I-DATAPROPERTY. At the logical layer, we represent the constructs as illustrated in Fig. 9 (schema-level constructs) and Fig. 10 (instance-level constructs). The figure provides the relational implementation of corresponding constructs in the meta-model. For the sake of simplicity, we omitted some attributes.

Each table at instance-level presents the references to the corresponding entries at schema level. In the figure we used short names as *uriService* or *uriInterface* to simplify the notation. To import the semantic annotation we follow a similar procedure and we link the annotation with the corresponding WSDL. Fig. 11 shows how the semantic annotation associated with S_1 is imported in our system. At schema level we introduced two entries *Booking* and *City* in the CLASS table representing two ontological concepts and one entry *linkTO* in OBJECTPROPERTY, representing the reference annotation between a WSDL

Class

OID	name
oid1	Service
oid2	InterfaceOperation
oid3	Interface
oid4	InputMessage

DATAPROPERTY

OID	name	type	classOID
oid8	name_interface	xsd:string	oid3
oid9	name_service	xsd:string	oid1

OBJECTPROPERTY

OID	name	subjectOID	objectOID
oid5	has_operation	oid3	oid2
oid6	implements	oid1	oid3
oid7	has_message	oid2	oid4

Fig. 9. An example of relational implementation of schema-level constructs in the SeeVa storage system

Individual

OID	URI	ClassOID
ind1	uriService	oid1
ind2	uriOperation	oid2
ind3	uriInterface	oid3
ind4	uriMessage	oid4

I-DATAPROPERTY

OID	name	value	IndividualOID	PropertyOID
data1	name_service	reservationService	ind1	oid8
data2	name_interface	bookingInterface	ind3	oid9

I-OBJECTPROPERTY

OID	name	subjectIndOID	objectIndOID	PropertyOID
obj1	has_operation	ind3	ind2	oid5
obj2	implements	ind1	ind3	oid6
obj3	has_message	ind2	ind4	oid7

Fig. 10. An example of relational implementation of instance-level constructs in the SeeVa storage system

Class

OID	name
oid0	Thing
oid2	InterfaceOperation
oid4	InputMessage
oid11	Booking
oid12	City

Individual

OID	URI	ClassOID
ind2	uriOperation	oid2
ind4	uriMessage	oid4
ind5	uriAnnotation1	oid11
ind6	uriAnnotation2	oid12

OBJECTPROPERTY

OID	name	subjectOID	objectOID
oid7	has_message	oid2	oid4
oid13	linkTO	oid0	oid0

I-OBJECTPROPERTY

OID	name	subjectIndOID	objectIndOID	PropertyOID
obj3	has_message	ind2	ind4	oid7
obj4	linkTO	ind2	ind5	oid13
obj5	linkTO	ind4	ind6	oid13

Fig. 11. An example of relational implementation of semantic annotations in the SeeVa storage system

component and a concept. In this case we introduced the entry *Thing* because the domain and range of *linkTO* are generic (unknown). At instance level we have two entries in the INDIVIDUAL table identified by OIDs ind5 and ind6, representing instances (e.g., annotations) of *Booking* and *City*, respectively, and two entries in I-OBJECTPROPERTY identified by OIDs obj4 and obj5, representing instances of *linkTO*.

4   Model-Based Service Discovery

Different kinds of match between the Web service request and Web service advertisements are checked according to seven steps, that are described in the following. Hereafter, we consider semantic Web service descriptions with only one operation. At the end of this section we will discuss how to extend the approach to semantic Web service descriptions with more than one operation.

Coarse-grained Web Service Filtering. This step is performed by relying on the set $C_{\mathcal{S}}^{op}$ of each advertisement $\mathcal{S}$ and the set $C_{\mathcal{R}}$ of the Web service request $\mathcal{R}$ as defined in Equation (2); this step filters out advertisements $\mathcal{S}$ that do not satisfy the following condition: $\forall c_{\mathcal{R}} \in C_{\mathcal{R}} \ \exists c_{\mathcal{S}} \in C_{\mathcal{S}}^{op}$ such that $c_{\mathcal{S}} \equiv c_{\mathcal{R}}$ or $c_{\mathcal{S}} \sqsubseteq c_{\mathcal{R}}$. This ensures that only Web service advertisements that provide operations/capabilities/profiles that are classified using the same or more specific concepts compared with the ones specified in the request are considered. Among them, more in-depth matching analysis based on the other kinds of match is checked. In the running example, $op_{\mathcal{S}_1}$ satisfies this condition (since $C_{\mathcal{S}_1}^{op} = C_{\mathcal{R}}$), as well as $op_{\mathcal{S}_2}$ (since $C_{\mathcal{S}_2}^{op}$ is not specified) and $op_{\mathcal{S}_3}$ (since MotelBooking $\sqsubseteq$ Booking).

Input Compatibility Checking. The input parameters of each Web service advertisement that has not been filtered out during the previous step are analysed in order to state if they are compatible with the inputs specified in the request $\mathcal{R}$; compatibility in this case is recognized iff the following condition holds: $\forall in_{\mathcal{S}} \in IN_{\mathcal{S}}^{op} \ \exists in_{\mathcal{R}} \in IN_{\mathcal{R}}$ such that $in_{\mathcal{R}} \equiv in_{\mathcal{S}}$ or $in_{\mathcal{R}} \sqsubseteq in_{\mathcal{S}}$. This ensures that the inputs specified within the request $\mathcal{R}$ are enough to execute $\mathcal{S}$. We distinguish among *strong input compatibility* (when $in_{\mathcal{R}} \equiv in_{\mathcal{S}}$ holds) and *weak input compatibility* (when $in_{\mathcal{R}} \sqsubseteq in_{\mathcal{S}}$ holds) of $\mathcal{S}$ with respect to $\mathcal{R}$. For instance, $\mathcal{S}_1$ and $\mathcal{S}_2$ present a strong input compatibility with respect to $\mathcal{R}$ since $IN_{\mathcal{S}_1}^{op} = IN_{\mathcal{S}_2}^{op} = IN_{\mathcal{R}}$; $\mathcal{S}_3$ does not present any kind of input compatibility, since there is no a corresponding input in $IN_{\mathcal{R}}$ for the concept Room $\in IN_{\mathcal{S}_3}^{op}$.

Exact Match Evaluation. EXACT match between the Web service request and a Web service advertisement $\mathcal{S}$ ensures that $\mathcal{S}$ and $\mathcal{R}$ represent exactly the same service. An exact match is recognized iff $\mathcal{S}$ and $\mathcal{R}$ present a strong input compatibility and the following condition holds: $\forall out_{\mathcal{R}} \in OUT_{\mathcal{R}} \ \exists out_{\mathcal{S}} \in OUT_{\mathcal{S}}^{op}$ such that $out_{\mathcal{S}} \equiv out_{\mathcal{R}}$. This condition is very restrictive. For instance, in the running example no EXACT matches can be recognized. This ensures high precision, but EXACT match constraints must be relaxed to include more search results, thus increasing the recall of the discovery process.

Plug-in Match Evaluation. A PLUG-IN match between S and R is recognized iff they present an input compatibility and the following condition holds: $\forall out_R \in OUT_R \ \exists out_S \in OUT_S^{op}$ such that $out_S \sqsubseteq out_R$. This kind of match derives from the software engineering domain, where plug-in is used to check if a given software component (the request) can be substituted by another component (the advertisement). For instance, the service S_1 presents a PLUG-IN match with the request R since, among the outputs in $OUT_{S_1}^{op}$, LuxuryHotel$\sqsubseteq$Hotel and Sauna$\equiv$Sauna. This still ensures high precision since all the requirements are satisfied, but also includes additional matching results, thus increasing the recall.

Relaxed Match Evaluation. A RELAXED match between S and R is recognized iff they present an input compatibility and the following condition holds: $\forall out_R \in OUT_R \ \exists out_S \in OUT_S^{op}$ such that $out_S \sqsubseteq out_R$ or out_R is a direct less specific concept compared with out_S, that is $out_R \sqsubseteq out_S$, but there is only one subclass relationship between them. In our approach, we rank different advertisements S that present a RELAXED match with R by defining a match degree $\text{RELAXED}_{deg}(S, R)$ as follows:

$$\text{RELAXED}_{deg}(S, R) = \frac{\sum_{i,j} Sim_H(out_S^i, out_R^j)}{|OUT_R|} \in [0, 1] \tag{4}$$

where i, j are indexes ranging over the set of elements in OUT_S^{op} and OUT_R, respectively, $|\cdot|$ denotes set cardinality and pairs $\langle out_S, out_R \rangle$ are the ones used to check the RELAXED match. The similarity measure $Sim_H \in [0, 1]$ between two concepts c_1 and c_2 according to a given concept hierarchy is defined as follows:

$$Sim_H(c_1, c_2) = \begin{cases} 1 & \text{if } c_1 = c_2 \\ 0.8^L & \text{if there are } L \text{ subClass relations} \\ & \text{between } c_1 \text{ and } c_2 \\ 0 & \text{otherwise} \end{cases} \tag{5}$$

The value 0.8 has been proved to be optimal in our past experiments on concept affinity [2]. For instance, the service S_2 presents a RELAXED match with the request R since, among the outputs in $OUT_{S_2}^{op}$, FinnishSauna$\sqsubseteq$Sauna ($Sim_H() = 0.8$) and Accomodation$\sqsupseteq$Hotel ($Sim_H() = 0.8$), therefore $\text{RELAXED}_{deg}(S_2, R) = (0.8 + 0.8)/2 = 0.8$.

Partial Match Evaluation. A PARTIAL match may occur if: (i) neither a strong nor a weak input compatibility holds; (ii) there exists a requested output $out_R \in OUT_R$ such that $\forall out_S \in OUT_S^{op} \ out_S \sqcap out_R \equiv \bot$; (iii) both (i) and (ii) are satisfied. If an equivalence or a path of subClass relationships cannot be found between two concepts, there can be several reasons: (i) the two concepts may have only a common ancestor within the ontology (for instance, Hotel and Lodge); (ii) they can be related concepts, but defined in different ontologies; (iii) the two concepts are not related at all. In the first case, the following similarity measure Sim_{CA} is adopted:

$$Sim_{CA}(c_1, c_2) = 0.8^{L_1+L_2} \tag{6}$$

where L_1 (resp., L_2) is the number of subClass relations between c_1 (resp., c_2) and the common ancestor. To check if two concepts are related, although they are defined in different ontologies, we rely on term similarity techniques based on WordNet, that have been successfully applied in the context of Web service discovery, as extensively described in [2]. We use the names of concepts as terms. In WordNet, the meaning of terms is defined by means of synsets. Synsets are related by eighteen different kinds of relationships. In particular, *hyponymy/hypernymy relations* are used to represent the specialization/generalization relationship between two terms. For instance, the concept Inn is not defined within the reference ontology in our running example, but, within WordNet, it is among the synonyms of the Lodge term and it is among the direct hyponyms of the Hotel term. This means that there is a term similarity, denoted as Sim_t, both between Inn and Hotel and between Inn and Lodge. The similarity between two terms corresponding to the names of concepts c_1 and c_2 is maximum (that is, equal to 1.0) if the terms belong to the same synset or coincide; otherwise, if they belong to different synsets, a path of hyponymy/hypernymy relations that connects the two synsets is searched: the highest the number of relationships in this path, the lowest the term similarity. The overall PARTIAL match degree $PARTIAL_{deg}(\mathcal{S}, \mathcal{R}) \in [0, 1]$ is computed as follows

$$PARTIAL_{deg}(\mathcal{S}, \mathcal{R}) = \frac{1}{2} \frac{\sum_{i,j} Sim(out_{\mathcal{S}}^i, out_{\mathcal{R}}^j)}{|OUT_{\mathcal{R}}|} + \frac{1}{2} \frac{\sum_{h,k} Sim(in_{\mathcal{R}}^h, in_{\mathcal{S}}^k)}{|IN_{\mathcal{S}}^{op}|} \tag{7}$$

where i, j are indexes ranging over the sets of elements in $OUT_{\mathcal{S}}^{op}$ and $OUT_{\mathcal{R}}$, respectively, h, k are indexes ranging over the sets of elements in $IN_{\mathcal{R}}$ and $IN_{\mathcal{S}}^{op}$, respectively, $|\cdot|$ denotes the set cardinality and $Sim(c_1, c_2)$ corresponds to one of the similarity measures among: (i) Sim_H, if $c_1 \sqsubseteq c_2$ within the same ontology; (ii) Sim_{CA}, if c_1 and c_2 presents a common ancestors within the same ontology; (iii) Sim_t, if c_1 and c_2 do not belong to the same ontology, but the terms corresponding to their names are related within WordNet. For instance, the service $\mathcal{S}_3$ presents a PARTIAL match with respect to $\mathcal{R}$ since, among the inputs within $IN_{\mathcal{S}_3}^{op}$, Room is not provided in $\mathcal{R}$ (that is, there is no input compatibility between $\mathcal{R}$ and $\mathcal{S}_3$). Moreover, if we consider $OUT_{\mathcal{R}}$ and $OUT_{\mathcal{S}_3}^{op}$, Hotel and Lodge have a common ancestor within the reference ontology, that is, $Sim_{CA}(\text{HOTEL}, \text{LODGE}) = 0.8^2 = 0.64$, therefore $PARTIAL_{deg}(\mathcal{S}_3, \mathcal{R}) = 1/2 \cdot [1.0/2] + 1/2 \cdot [0.64/1] = 0.57$. RELAXED and PARTIAL matches constitute a relaxation compared with the first two kinds of match since, when a required output is not found, a more generic or semantically related output is searched for within the Web service advertisement. In particular, in the PARTIAL match evaluation, also the input compatibility check is relaxed. This ensures better recall, while at the same time the precision is only slightly negatively affected, thanks to the ontology-based computation of Sim_H and the WordNet-based computation of Sim_{CA}, as confirmed by experimental results.

Ranking and Filtering. The kinds of match explained above present a decreasing level of precision and an increasing level of recall since each of them corresponds to a relaxation of constraints compared with the previous one. Therefore, we defined a ranking among them according to the ranking function $\prec_m$, where if $m_1 \prec_m m_2$ then m_1 is ranked better than m_2:

$$\text{EXACT} \prec_m \text{PLUG-IN} \prec_m \text{RELAXED} \prec_m \text{PARTIAL}$$

Within the last two kinds of match, ranking is based on RELAXED_{deg} and PARTIAL_{deg} evaluation, while no ranking is required among EXACT and PLUG-IN matches, since they all represent search results that completely satisfy the request. Finally, a threshold-based filtering is performed, by including among search results all EXACT and PLUG-IN matches, all RELAXED matches such as $\text{RELAXED}_{deg} \geq \delta$ and PARTIAL matches such as $\text{PARTIAL}_{deg} \geq \delta$, where δ is the threshold set by the requester.

Discussion. Till now, we considered Web service advertisements S_i with only one operation. If there are S_i with more operations, the goal becomes to find a Web service advertisement that is able to provide, in one of its operations, the required $IN_\mathcal{R}$ and $OUT_\mathcal{R}$. In this case, S_i is split into several advertisements (one for each operation) and the discovery procedure is performed in the same way (that is, additional operations provided by S_i are ignored). This is the same rationale behind match evaluation for what concerns requested outputs. If an advertisement S_i presents more operations that match against $\mathcal{R}$, then the best one (according to the ranking step) is chosen.

4.1   The Hybrid Matchmaking Algorithm

The hybrid matchmaking algorithm we designed relies on the Roman principle of *divide et impera* to enable an iterative reduction of the search space Σ (identified as the set of Web service advertisements that can potentially match the request $\mathcal{R}$) in order to accelerate the Web service discovery procedure. The pseudo-code of the algorithm is shown in Algorithm 1. In the algorithm, an advertisement is featured by three properties: the kind of input compatibility among strong, weak or none ($S_i.IN\_comp$, see row 5); the similarity degree $S_i.simDegree$ (see row 18); the kind of match $S_i.match$ (see row 24).

Firstly, on row 1 some buffer variables are initialized: the set of discovery results (Σ'), the sets that will contain the advertisements with a strong and a weak input compatibility wrt the request (Σ^{STRONG} and Σ^{WEAK}, respectively), the set that will contain the partial matches (Σ^{PARTIAL}). The instruction on row 2 concerns the exclusion from the search space Σ of all the advertisements that have been filtered out during the *coarse-grained Web service filtering* step. A second step concerns the input compatibility checking (rows 3-12). For each advertisement in Σ, it is checked if $|IN_S^{op}| \leq |IN_\mathcal{R}|$: if it is FALSE, neither the strong nor the weak input compatibility conditions can be satisfied; otherwise the input compatibility conditions are checked and the sets Σ^{STRONG} and Σ^{WEAK} are properly populated.

Algorithm 1. Hybrid matchmaking algorithm

Input : The set of Web service advertisements Σ, a Web service request $\mathcal{R}$, a similarity threshold δ.

Output: The ranked set Σ' of advertisements that match against $\mathcal{R}$.

1   $\Sigma' \leftarrow \emptyset$; $\Sigma^{\text{STRONG}} \leftarrow \emptyset$; $\Sigma^{\text{WEAK}} \leftarrow \emptyset$; $\Sigma^{\text{PARTIAL}} \leftarrow \emptyset$;

2   $\Sigma \leftarrow \text{COARSEGRAINEDFILTER}(\Sigma, \mathcal{R})$;

3   **foreach** $\mathcal{S}_i \in \Sigma$ **do**

4   **if** $(|IN^{op}_{\mathcal{S}_i}| \leq |IN_{\mathcal{R}}|)$ **then**

5         $\mathcal{S}_i.IN\_comp \leftarrow \text{CHECKINPUTCOMP}(\mathcal{S}_i, \mathcal{R})$;

6   **if** $\mathcal{S}_i.IN_comp == $ '**strong**' **then** $\Sigma^{\text{STRONG}} \leftarrow \Sigma^{\text{STRONG}} \cup \{\mathcal{S}_i\}$;

7   **else**

8   **if** $\mathcal{S}_i.IN_comp == $ '**weak**' **then** $\Sigma^{\text{WEAK}} \leftarrow \Sigma^{\text{WEAK}} \cup \{\mathcal{S}_i\}$;

9   **else** $\mathcal{S}_i.IN\_comp \leftarrow$ '**none**';

10 Compute Σ_{EXACT} on Σ^{STRONG};

11 $\Sigma' \leftarrow \Sigma' \cup \Sigma_{\text{EXACT}}$;

12 Compute $\Sigma_{\text{PLUG-IN}}$ and Σ_{RELAXED} on $\Sigma^{\text{STRONG}} \cup \Sigma^{\text{WEAK}}$;

13 $\Sigma' \leftarrow \Sigma' \cup \Sigma_{\text{PLUG-IN}}$;

14 **foreach** $\mathcal{S}_i \in \Sigma_{\text{RELAXED}}$ **do** $\mathcal{S}_i.simDegree \leftarrow \text{RELAXED}_{deg}(\mathcal{S}_i, \mathcal{R})$;

15 Rank Σ_{RELAXED} wrt $\mathcal{S}_i.simDegree$;

16 $\Sigma' \leftarrow \Sigma' \cup \Sigma_{\text{RELAXED}}$;

17 **foreach** $\mathcal{S}_i \in \Sigma \setminus \Sigma'$ **do**

18   **if** $(min(|OUT^{op}_{\mathcal{S}}|, |OUT_{\mathcal{R}}|)/|OUT_{\mathcal{R}}| + min(|IN_{\mathcal{R}}|, |IN^{op}_{\mathcal{S}}|)/|IN^{op}_{\mathcal{S}}| \geq 2\delta)$ **then**

19   Check PARTIAL match between $\mathcal{S}_i$ and $\mathcal{R}$;

20   **if** $\mathcal{S}_i.match == $ '**partial**' **then**

21            $\mathcal{S}_i.simDegree \leftarrow \text{PARTIAL}_{deg}(\mathcal{S}_i, \mathcal{R})$;

22   **if** $\mathcal{S}_i.simDegree \geq \delta$ **then** $\Sigma^{\text{PARTIAL}} \leftarrow \Sigma^{\text{PARTIAL}} \cup \{\mathcal{S}_i\}$;

23 Rank Σ^{PARTIAL} wrt $\mathcal{S}_i.simDegree$;

24 $\Sigma' \leftarrow \Sigma' \cup \Sigma^{\text{PARTIAL}}$;

25 **return** Σ';

The advertisements that present an EXACT match with the request $\mathcal{R}$ are extracted as follows:

$$\Sigma_{\text{EXACT}} = \bigcap_{j=1}^{|OUT_{\mathcal{R}}|} \Sigma_{OUT}(out^j_{\mathcal{R}}) \tag{8}$$

where $\Sigma_{OUT}(out^j_{\mathcal{R}})$, with $j = 1, \dots |OUT_{\mathcal{R}}|$, denotes the advertisements that provide an output that has been semantically annotated with a concept $out^j_{\mathcal{R}} \in OUT_{\mathcal{R}}$ or an equivalent one. The Σ_{EXACT} set is computed on the advertisements that belong to the Σ^{STRONG} set only (rows 13-14). The advertisements that present a PLUG-IN match with the request $\mathcal{R}$ are extracted as follows:

$$\Sigma_{\text{PLUG-IN}} = \left[\bigcap_{j=1}^{|OUT_{\mathcal{R}}|} (\Sigma_{OUT}(out_{\mathcal{R}}^j) \cup \widehat{\Sigma}_{OUT}(out_{\mathcal{R}}^j)) \right] - \Sigma_{\text{EXACT}} \qquad (9)$$

where $\widehat{\Sigma}_{OUT}(out_{\mathcal{R}}^j)$, with $j = 1, \dots |OUT_{\mathcal{R}}|$, denotes the advertisements that provide an output that has been semantically annotated with a concept that is more specific than $out_{\mathcal{R}}^j \in OUT_{\mathcal{R}}$. The advertisements that present a RELAXED match with the request $\mathcal{R}$ are extracted as follows:

$$\Sigma_{\text{RELAXED}} = [\bigcap_{j=1}^{|OUT_{\mathcal{R}}|} (\Sigma_{OUT}(out_{\mathcal{R}}^j) \cup$$
$$\widehat{\Sigma}_{OUT}(out_{\mathcal{R}}^j) \cup \overline{\Sigma}_{OUT}(out_{\mathcal{R}}^j))] \qquad (10)$$
$$- \Sigma_{\text{PLUG-IN}}$$

where $\overline{\Sigma}_{OUT}(out_{\mathcal{R}}^j)$, with $j = 1, \dots |OUT_{\mathcal{R}}|$, denotes the advertisements that provide an output that has been semantically annotated with a concept that is more generic than $out_{\mathcal{R}}^j \in OUT_{\mathcal{R}}$. The sets $\Sigma_{\text{PLUG-IN}}$ and Σ_{RELAXED} are computed on the advertisements that belong to the Σ^{STRONG} and Σ^{WEAK} sets (rows 15-20). In the running example, $\Sigma_{OUT}(\text{HOTEL}) = \emptyset$, $\widehat{\Sigma}_{OUT}(\text{HOTEL}) = \{S_1\}$, $\overline{\Sigma}_{OUT}(\text{HOTEL}) = \{S_2\}$, $\Sigma_{OUT}(\text{SAUNA}) = \{S_1\}$, $\widehat{\Sigma}_{OUT}(\text{SAUNA}) = \{S_2\}$, $\overline{\Sigma}_{OUT}$ $(\text{SAUNA}) = \emptyset$. Therefore, $\Sigma_{\text{PLUG-IN}} = \{S_1\} \cap \{S_1, S_2\} = \{S_1\}$ and $\Sigma_{\text{RELAXED}} = \{S_1, S_2\} \cap \{S_1, S_2\} - \{S_1\} = \{S_2\}$.

Rows 21-29 show the partial match evaluation. Partial match evaluation is not performed if the condition on row 22 is not satisfied. Consider $\mathcal{R}$ and S_3 in the running example. If the threshold is $\delta = 0.8$, then the maximum possible value for PARTIAL_{deg} is the one expressed in the condition on row 22, that is, 0.75. This value can be obtained if all the $Sim(c_1, c_2)$ values in the PARTIAL_{deg} are equal to 1.0. In this example, whatever are the IO of $\mathcal{R}$ and S_3, PARTIAL_{deg} will be always below $\delta = 0.8$ and S_3 will be excluded from search results (that is, PARTIAL match evaluation can be avoided). The strong points of the algorithm concern the instructions introduced to reduce the search space and speed up the discovery procedure (see rows 2, 4, 13, 15, 22). Moreover, ranking and filtering of results with respect to the similarity degree is always performed within a single kind of match (see rows 19 and 28), thus reducing the computation time, as confirmed by experimental results.

4.2   Exploitation of the SeeVa Storage System

Datalog$^\pm$ rules. Reasoning and querying in NYAYA are based on Datalog$^\pm$ [4,5], a family of rule-based languages that extends Datalog [6]. In Datalog$^\pm$ language, Datalog is extended by allowing features such as existential quantifiers, the equality predicate, and the truth constant false to appear in rule heads. At the same time, the resulting language is syntactically restricted, so as to achieve decidability and in some relevant cases even tractability. This features are required for an effective and efficient reasoning over data. Web Service matchmaking is performed on top of NYAYA through the formulation of Datalog$^\pm$

rules that use predicates, classes and roles presented in the previous section. In Datalog$^{\pm}$ rules, a unary predicate corresponds to each class and a binary predicate corresponds to each role. For instance, the atom $Service(X)$ corresponds to the class $Service$ and the atom $implements(X, Y)$ corresponds to the role $implements$. Rules are combined to form Datalog$^{\pm}$ programs, which check the different kinds of match. In particular, referring to the semantic representation of the WSDL Schema elements in Fig. 7, the subset $\Sigma_{OUT}(out_{\mathcal{R}}^j)$ is extracted as follows:

OUT($?x,out_{\mathcal{S}_R}^i$) :- Service($?x$), implements($?x,?y$), Interface($?y$),
    has_operation($?y,?z$), interfaceOperation($?z$), has_message($?z,?o$),
    outputMessage($?o$), elementContent_message($?o,?e$), Element($?e$),
    linkTO($?e,out_{\mathcal{S}_R}^i$)

The subset $\widehat{\Sigma}_{OUT}(out_{\mathcal{R}}^j)$ is obtained as follows:

$\widehat{OP}$($?x,op$) :- OP($?x,op$)
$\widehat{OP}$($?x,op$) :- $\widehat{OP}$($?x,?y$), subClass($?y,op$)

The subset $\overline{\Sigma}_{OUT}(out_{\mathcal{R}}^j)$ is obtained as follows:

$\widehat{OP}$($?x,op$) :- OP($?x,op$)
$\widehat{OP}$($?x,op$) :- $\widehat{OP}$($?x,?y$), subClass($?y,op$)

Finally, the subset $\Sigma_C(c_{\mathcal{R}}^n)$ is computed as follows:

OP($?x,op_{\mathcal{S}_R}$) :- Service($?x$), implements($?x,?y$), Interface($?y$),
    has_operation($?y,?z$), InterfaceOperation($?z$),
    linkTO($?z,op_{\mathcal{S}_R}$)

Translation of Datalog$^{\pm}$ rules into storage programs. The Datalog$^{\pm}$ rules are formulated as queries on the database. Therefore NYAYA provides a set of *storage-programs* to map the entities involved in the query into the logical representation of the database. For the sake of clearness, in the following example programs we express, for each predicate, the name of attributes, their values (constant or variable). Moreover we use the prefix TMP to indicate those predicates that do not get materialized into the database, since at the moment we do not want to incur into view maintenance problems. For instance the following storage program is associated with the Datalog$^{\pm}$ rule corresponding to the subset $\Sigma_C(c_{\mathcal{R}}^n)$:

```
TMP_OP(S: I1, op: C2) :- Class(OID: C1, Name: 'Service'),
        Individual(OID: I1, Uri: U1, ClassOID: C1),
        ObjectProperty(OID: O1, Name: 'implements'),
        i-ObjectProperty(SubjIndOID: I1, ObjIndOID: I2,
                        PropertyOID: O1),
        ObjectProperty(OID: O2,
                        Name: 'has_interfaceOperation'),
        i-ObjectProperty(SubjIndOID: I2, ObjIndOID: I3,
                        PropertyOID: O2),
        ObjectProperty(OID: O3, Name: 'linkTO'),
```

```
i-ObjectProperty(SubjIndOID: I3, ObjIndOID: I4,
                 PropertyOID: O3),
Individual(OID: I4, ClassOID: C2),
Class(OID: C2, name: N)
```

Similarly we write the storage programs for the other Datalog$^\pm$ rules producing the temporary predicate TMP_OUT with attributes S and out (such as we produce the temporary predicate TMP_IN with attributes S and in). Therefore, given Out1, ..., Outn output names and IN1, ..., INh input names, we write the storage programs of the EXACT match rule r1 as follows:

```
TMP_EXACT(S: Y) :- TMP_OP(S: Y, op: N),
                   TMP_OUT(S: Y, out: W1), ...,
                   TMP_OUT(S: Y, out: Wn),
                   TMP_IN(S: Y, out: Z1), ...,
                   TMP_IN(S: Y, out: Zh)
```

where W1, ..., Wn (resp., Z1, ..., Zh) are OIDs in the table Class corresponding to Out1, ..., Outn (resp., IN1, ..., INh) values for the attribute name. Referring to the example in the previous sections, we can rewrite r1 using the value Booking for N, Accomodation and Activity as Out1, ..., Outn and City as IN1, ..., INh.

Finally we introduce a set of *reasoning-level programs* to provide the reasoning capabilities: the result is the final Datalog$^\pm$ program executed by NYAYA. The *reasoning-level program* changes across different ontological formalisms. Currently, we exploit the inference rules for *subsumption* capabilities in $\mathcal{DL} - \mathcal{L}ite$ ontologies. For instance, referring to the previous section, the program p1 is written as follows:

```
TMP_OP_V(S: I1, op: C2) :- TMP_OP(S: I1, op: C2)
TMP_OP_V(S: I1, op: C3) :- TMP_OP(S: I1, op: C2),
              SubClass(ClassOID: C2, SubClassOID: C3)
```

where, if we want to instantiate p1, C2 corresponds to the OID in Class, where the attribute name has value Booking. We can write in the same way the PLUGIN rule r4 and the PARTIAL program p2.

5   System Validation

5.1   Implementation Issues

The architecture of the SeeVa framework is shown in Fig. 12. It is based on the NYAYA architecture described in [9]. The main components are the *Service Manager*, the *Importer*, the *Storage System*, the *Reasoner* and the *MatchMaker*.

The *Service Manager* (SM) is in charge of capturing the collection of semantic Web service descriptions. Both the OWLS-MX [21] and the SAWSDL-MX [22] matchmakers expose prototype tools to extract service descriptions both from the WSDL and the corresponding semantic annotation. We imported the outputs of such tools into the SeeVa framework. SM also presents a *Conformer*

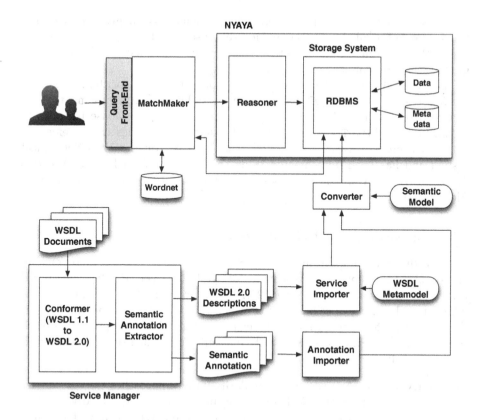

Fig. 12. The SeeVa framework architecture

to standardize the WSDL 1.1 representation into 2.0[3]. Then a *Semantic Annotation Extractor* extracts the semantic annotation associated with the service description. We consider RDF, RDFS, and OWL as annotation models. However our approach is extensible to other ontologies.

A *Service Importer* (SI) and an *Annotation Importer* (AI) parse the WSDL 2.0 document and the corresponding semantic annotation, respectively. The former extracts elements and attributes from the WSDL document through our metamodel describing the *constructs* of interest in the WSDL schema[4]. The latter extracts the TBox (terminological) axioms and the ABox (assertional) facts from the annotation.

The *Storage System* of NYAYA populates a relational database with the incoming WSDL elements (data) and semantic annotations (meta-data). The *Converter* is in charge of associating each construct of the source model to the corresponding

---

[3] The *Conformer* exploits the converter available at http://www.w3.org/2006/02/WSDLConvert.html

[4] http://www.w3.org/TR/2006/CR-wsdl20-primer-20060327/wsdl20-primer-diff.html

primitive of the meta-model and storing the elements accordingly. The *Reasoner* is composed of a rule-based engine and a basic set of inference rules defined by means of a *Rule Manager* that dictates how new information can be inferred from ground facts and ontological domain knowledge.

The *MatchMaker* implements the service matchmaking algorithm. It exploits the *Reasoner* to perform logic-based reasoning and it also extracts semantic Web service descriptions and annotations from the relational database to evaluate RELAXED_{deg} and PARTIAL_{deg} evaluation. The latter capabilities are implemented within a *Similarity Evaluator* module, that is based on the WordNet lexical system and has been implemented using the WordNet Java library. A *Query Front-End* allows Web service requesters to submit Datalog$^{\pm}$ programs in a user friendly way[5].

5.2   Comparison with Related Work

Table 1 shows a comparison between `SeeVa` and some service matchmaking approaches, that represent the features of similarity-based [13,28], logic-based [15,17,24] and hybrid ones [2,12,19,21,22]. The comparison is based on: (i) the different Web service specifications to which the approach can be applied; (ii) the matching information that is returned by the matchmaking procedure (such as the kind of match and the similarity measure); (iii) when the similarity evaluation is applied (if any).

Similarity-Based Approaches. Similarity-based approaches do not classify discovery results using the kinds of match, but provide a quantitative evaluation of the *similarity degree* between Web service request and advertisements, that is, a numeric value computed by applying different techniques (data mining techniques, structured graph matching, concept distance over ontologies/thesauri, content-based information retrieval metrics, etc.) to the elements in Web service descriptions. These approaches are usually not based on semantic Web service descriptions, but only applied to WSDL: WOOGLE (Web Service Search Engine) [13] uses a clustering algorithm for identifying relationships among the terms used within WSDL and evaluates operations, input and output similarities based on the relationships; URBE (Uddi Registry By Example) [28] evaluates the similarity degree $fSim$ by taking into account the names of Web service elements (semantic analysis) and the number of operations and of their parameters (structural analysis). URBE also has a semantic extension, where concepts used to annotate operations and IO parameters according to the SAWSDL specification are considered. Other similarity-based approaches present the same features highlighted in Table 1 for [13,28]. Recent approaches [33,35] use clustering for Web service discovery. Authors in [35] apply techniques to group Web services by exploiting tags assigned by users. Tags are intended to provide additional contextual and semantic information compared to WSDL-based Web service description, although their semantics is not as explicit as for semantic descriptions based

---

[5] Look at www.nyaya.eu for a demo.

either on OWL-S, WSMO or SAWSDL. In [33] an information retrieval technique known as latent semantic analysis is applied to the collection of WSDL files, to cluster similar services, using WordNet as lexical database. No formal semantic tools, such as ontologies, are applied here. Such approaches assume that semantics of terms for Web service description is shared between service consumers and providers and their success relies on this assumtpion, that constitutes their main limitation.

Logic-Based Approaches. Logic-based approaches model both Web service request and advertisements as a set of logical expressions or constraints (e.g., restrictions on the Web service input/output class, expressed as a simple or complex ontological concept) [15–17,24]. In [17] the overall expression representing an OWL-S service profile is mapped into a single $\mathcal{SHIQ}(D)$ expression and DL-based deductive facilities are applied to check if a Web service advertisement satisfies all the request constraints and only them (EXACT match), satisfies all the constraints but adds additional restrictions (PLUG-IN match) or satisfies only a subset of the request constraints (PARTIAL match). Similarly, in [15] partial matches are checked investigating the compatibility between S and R in terms of concept intersection or non-disjointness ($S \sqcap R$) with respect to a knowledge base or ontology. Ontology-based service discovery approaches [7,29] can be classified among logic-based ones. These efforts usually refer to a single ontology, which can be a limitation in a decentralized environment, and are oriented towards precision maximization; therefore, they are characterized by low recall values.

Hybrid Approaches. Hybrid matchmaking approaches apply in a combined way logic-based and similarity-based techniques. In this way, they try to increase both precision and recall of search results. OWLS-MX [21], WSMO-MX [19], SAWSDL-MX [22] work in very similar ways: if logic-based comparison fails, an approximate match is introduced (called nearest-neighbor match within OWLS-MX and SAWSDL-MX or fuzzy similarity match within WSMO-MX) where service matchmaking is evaluated through different IR metrics (loss-of-information measure, extended Jacquard similarity coefficient, cosine similarity value, Jensen-Shannon information divergence-based similarity value). In MAMAS [12], when a constraint in the Web service request is not satisfied by any constraint in the Web service advertisement, a penalty is assigned. The higher is the total penalty, the lower is the compatibility between the request and the advertisement. In FC(Functional Compatibility)-MATCH [2], a DL-based Web service advertisement description S is extracted from SAWSDL specification and described through sets of concepts representing categories, operations, inputs and outputs. Matching degree is computed by considering both logic-based subsumption between concepts and the text similarity between concept names according to terminological relationships extracted from WordNet, thus enabling a de-facto comparison across different ontologies. The only approach that abstracts from the adopted specification is MDSM (Model-Driven Service Matchmaker) [23], which relies on existing hybrid solutions (OWLS-MX, WSMO-MX and SAWSDL-MX) and invokes them depending on the semantic Web

Table 1. Model-driven comparison of approaches for Web service matchmaking

Approach	Web service model	Kinds of match	Similarity evaluation
WOOGLE [13]	WSDL	-	Over the whole service
URBE [28]	WSDL/SAWSDL	-	Over the whole service
Grimm [15]	DL-based from OWL-S	Exact, Plug-in, Subsumes, Intersection, Disjoint	-
Horrocks and Li [17]	DL-based from OWL-S	Exact, Plug-in, Subsumes, Intersection, Disjoint	-
Lausen et al. [24]	WSMO	Exact, Plug-in, Subsumes, Intersect, Disjoint	-
FC-MATCH [2]	DL-based from SAWSDL	Exact, Plug-in, Subsumes, Intersection, Mismatch	For intersection match
MAMAS [12]	DL-based representation	Exact, Plug-in, Subsumes, Intersection, Disjoint	Penalties on the whole service
WSMO-MX [19]	WSMO	Equivalence, Plug-in, Inverse-plug-in, Intersection, Fuzzy similarity, Neutral, Disjunction	For fuzzy similarity match
OWLS-MX [21]	OWL-S	Exact, Plug-in, Subsumes, Subsumed-by, Nearest-neighbor, Fail	For nearest-neighbor match
SAWSDL-MX [22]	SAWSDL	Exact, Plug-in, Subsumes, Subsumed-by, Nearest-neighbor, Fail	For nearest-neighbor match
SeeVa	OWL-S/WSMO/SAWSDL	Exact, Plug-in, Relaxed, Partial, Fail	For Relaxed and Partial match

service description specification. To address Web service discovery across different specifications, approaches like the one described in [23] rely on existing language-specific solutions and invoke them depending on the specification used for representing both the Web service advertisements (i.e., the descriptions of available Web services) and the Web service request. Therefore, for each comparison they rely on the features and reasoning capabilities of each language-specific matchmaker. Compared to existing solutions, SeeVa presents a Web service model that integrates the features of existing semantic Web service specifications and provides an hybrid algorithm based on such a model where different similarity measures are applied depending on the kind of match and enables Web service comparison across different ontologies.

5.3   Experimental Results

We compared our framework against representative solutions among the related work. Specifically, we performed a comparison against: (i) the logic-based service matchmaker OWLS-M0 and its hybrid variant OWLS-MX [21], which uses the cosine similarity if logic-based comparisons fail; (ii) another hybrid service matchmaker, SAWSDL-MX [22], which computes the extended Jacquart similarity coefficient if logic-based comparisons fail; (iii) the similarity-based matchmaker URBE [28], considering the semantic extension that extracts Web service descriptions from their SAWSDL specifications.

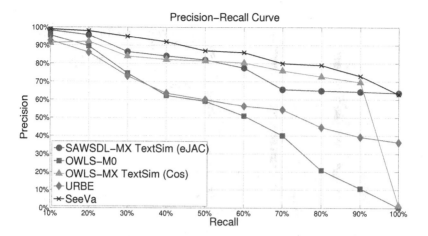

Fig. 13. Precision-Recall curve of the compared Web service matchmaking approaches

Datasets. For experimental comparison, we used the SME^2 Semantic Web Service Matchmaker Evaluation Environment v2.1[6]. Within the SME^2 environment, we used two public available datasets. The first one is *OWL-S Service Retrieval Test Collection* (OWLS-TC, version 4), a collection of 1083 OWL-S1.1 Web services. The second dataset is the *SAWSDL Service Retrieval Test Collection* (SAWSDL-TC, version 3), containing 1080 SAWSDL Web services. Each dataset contains 42 queries used to perform the experiments. For each dataset, queries have already been associated with relevant service advertisements, thus enabling the evaluation of precision/recall for each matchmaking algorithm. OWLS-M0 and OWLS-MX have been applied on the OWLS-TC4, SAWSDL-MX and URBE have been applied on SAWSDL-TC3. We applied our approach to the union set of the two datasets and we verified that SeeVa outperforms the other approaches in terms of precision and recall and query response time. All the experiments were performed on a dual quad core 2.66GHz Intel Xeon, running Linux Gentoo, with 8 GB of memory, 6 MB cache and a 2-disk 1Tbyte striped RAID array, and on PostgreSQL 8.3.

Matchmaking Effectiveness. *Precision P* (i.e., the ratio of the number of relevant retrieved Web services to the total number of retrieved Web services) and *recall R* (i.e., the ratio of the number of relevant retrieved Web services to the total number of relevant Web services in the dataset) have been adopted to evaluate the performances of compared approaches. For each level r_j of recall we calculated the average max precision of queries in $[r_j, r_{j+1}]$, i.e. $P(r_j) = \frac{1}{|Q|} \cdot \sum_{q \in Q} max_{r_j \leq r \leq r_{j+1}} P(r)$, where Q is the set of queries and we used $\kappa = 10$ recall levels within the $[0..1]$ range, that is, recall levels are at equidistant steps $\frac{i}{\kappa}$, where $i = 1 \ldots \kappa$. The graph in Fig. 13 shows the Precision-Recall curve for all the

---

[6] http://projects.semwebcentral.org/projects/sme2/

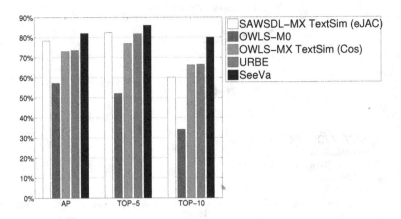

Fig. 14. Comparison between service matchmaking approaches with respect to the *TOP-n precision P^n* and the *Average Precision AP* metrics

compared matchmakers. The SeeVa framework shows good results compared to hybrid solutions such as OWLS-MX and SAWSDL-MX and compared to URBE. Good precision results of the URBE matchmaker are due to a semantic-based evaluation of service similarity, which increases the precision values compared to other kinds of similarity-based approaches, that perform text-based comparisons and suffer from the presence of synonyms within the descriptions of compared services, and makes URBE precision comparable to OWLS-M0 matchmaker. Hybrid solutions maintain high precision and high recall values since they are able to recognize relevant results also among matches that are neither Exact not Plug-in, due to the evaluation of similarity degree. Recall of SeeVa is maintained high also for high precision values, compared to the other hybrid solutions, thanks to the evaluation of PARTIAL$_{deg}$ among results that contain concepts represented within different ontologies (see Section 4).

Precision and recall do not take into account the ranking order in the result set. Therefore, we also computed the *TOP-n precision P^n* and the *Average Precision AP*. The *TOP-n precision P^n* is the average precision at a given cutoff point n, that is, the precision computed on the first n results

$$P^n = \frac{1}{|Q|} \sum_{q \in Q} \frac{|\Sigma_q^n \cap \mathcal{R}_q|}{n} \tag{11}$$

where Σ_q^n is the set of first n returned services for query q and $\mathcal{R}_q$ is the set of relevant services in the dataset for query q. The *Average Precision AP* is defined as follows

$$AP = \frac{1}{|Q|} \sum_{q \in Q} \left\{ \frac{1}{|\mathcal{R}_q|} |\Sigma_{r=1}^{|\Sigma_q|} isrel(r) \frac{count(r)}{r} \right\} \tag{12}$$

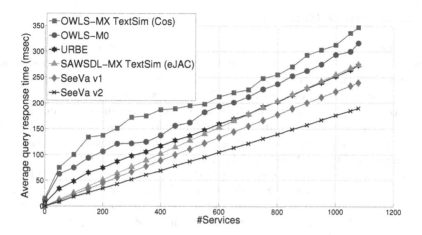

Fig. 15. Comparison service matchmaking approaches with respect to the average response time

where Σ_q is the set of returned services for query q, $isrel(r) = 1$ if the result at rank r is relevant and 0 otherwise and $count(r) = \sum_{i=1}^{r} isrel(i)$. Fig. 14 shows the AP, P^5 and P^{10} values for the compared matchmakers. SeeVa again outperforms the other solutions thanks to the differentiation between RELAXED$_{deg}$ and PARTIAL$_{deg}$, that is not performed by the other hybrid solutions.

Performance Evaluation. Fig. 15 shows the average query response time (msec) with respect to the increasing number of services to test the scalability of the systems. The goal of this experiment is to demonstrate how our approach does not pay its increased precision and recall in terms of worse query response time, but its response time is in line with the other ones and does not significantly worsen as the number of compared services increases. To evaluate the response time, we submitted all the queries in the datasets to the matchmakers. For each submission, we calculated the time needed to finish the execution of the matchmaking procedure (we did not consider preliminary operations of matchmakers to load Web service descriptions, see for details the SME environment). We performed each evaluation ten times and we computed the average time. In all cases, the maximal deviation from the average was no more than 3%. The results are very promising. For each kind of match the SeeVa average query response time is always in the range [0, 200] msec. This is due both to the internal data organization of our system and, in particular, to the optimization heuristics implemented within the matchmaking algorithm (see Section 4.1). SeeVa outperforms the other matchmakers. In particular, it outperforms SeeVa v1, the preliminary version of the framework proposed in [8], where we did not use the storage system introduced here and we did not implement any optimization heuristics in the matchmaking algorithm.

6   Concluding Remarks

The adoption of different SWS specifications impacts upon the scope of the Web service discovery process. In this paper we propose a SWS model that abstracts the elements of all the most common specifications and we provide an efficient representation of the model within `SeeVa`, a storage system based on a Datalog engine that enables language-independent reasoning capabilities. The SWS model and its representation within `SeeVa` are the starting point for a new hybrid service matchmaking algorithm, where we implemented new mechanisms to speed up and improve the retrieval of relevant Web services that match a given Web service request. Experimental results demonstrated the accuracy of the proposal, in terms of precision and recall, and its efficiency. Following existing service matchmakers, we did not include pre- and post-conditions within the model. This choice has also been influenced by the lack of pre- and post-conditions examples in the available benchmarks and datasets. These aspects must be investigated as future extension of the model. Moreover, since the model has been studied for Web service discovery purposes, its application for Web service composition deserves additional research. Finally, Quality of Service (QoS) is another orthogonal perspective that should be considered after a functionality-based service selection [18]. The integration of all these aspects is crucial to enable testing of our approach in real world applications of semantic Web services, spanning from Smart Grids [31] to service use in Industry 4.0 [25].

References

1. Atzeni, P., Cappellari, P., Torlone, R., Bernstein, P.A., Gianforme, G.: Model-independent schema translation. VLDB Journal **17**(6), 1347–1370 (2008)
2. Bianchini, D., Antonellis, V.D., Melchiori, M.: Flexible Semantic-based Service Matchmaking and Discovery. World Wide Web Journal **11**(2), 227–251 (2008)
3. Bussler, C., de Bruijn, J., Feier, C., Fensel, D., Keller, U., Lara, R., Lausen, H., Polleres, A., Roman, D., Stollberg, M.: Web Service Modeling Ontology. Applied Ontology **1**(1), 77–106 (2005)
4. Calì, A., Gottlob, G., Lukasiewicz, T.: A General Datalog-Based Framework for Tractable Query Answering over Ontologies. In: PODS. pp. 77–86 (2009)
5. Calì, A., Gottlob, G., Pieris, A.: Advanced processing for ontological queries. In: VLDB. pp. 554–565 (2010)
6. Ceri, S., Gottlob, G., Tanca, L.: What you always wanted to know about Datalog (and never dared to ask). IEEE TKDE **1**(1), 146–166 (1989)
7. Chhun, S., Moalla, N., Ouzrout, Y.: Ontology-based approaches for semantic service selection in business process re-engineering. In: Enterprise Interoperability VI. pp. 63–73 (2014)
8. De Virgilio, R., Bianchini, D.: A metamodel approach to flexible semantic web service discovery. In: CIKM. pp. 1309–1312 (2010)
9. De Virgilio, R., Nostro, P.D., Gianforme, G., Paolozzi, S.: A scalable and extensible framework for query answering over RDF. World Wide Web Journal **14**, 599–622 (2011)
10. De Virgilio, R., Orsi, G., Tanca, L., Torlone, R.: Semantic data markets: a flexible environment for knowledge management. In: CIKM. pp. 1559–1564 (2011)

11. De Virgilio, R., Orsi, G., Tanca, L., Torlone, R.: Nyaya: A system supporting the uniform management of large sets of semantic data. In: ICDE. pp. 1309–1312 (2012)
12. Di Sciascio, E., Di Noia, T., Donini, F.: Semantic Matchmaking as Non-Monotonic Reasoning: A Description Logic Approach. Journal of Artificial Intelligence Research **29**, 269–307 (2007)
13. Dong, X., Halevy, A.Y., Madhavan, J., Nemes, E., Zhang, J.: Similarity Search for Web Services. In: VLDB. pp. 372–383. Toronto, Canada (2004)
14. Farrell, J., Lausen, H.: Semantic Annotations for WSDL and XML Schema. Tech. rep., W3C (2007)
15. Grimm, S.: Semantic Web Services: Concepts, Technologies, and Applications, chap. Discovery: Identifying Relevant Services, pp. 211–244. Springer (2007)
16. Hobold, G., Siqueira, F.: Discovery of Semantic Web Services compositions based on SAWSDL annotations. In: IEEE 19th Int. Conference on Web services (2012)
17. Horrocks, I., Li, L.: A Software Framework for Matchmaking Based on Semantic Web Technology. Int. Journal of Electronic Commerce (IJEC) **8**(4), 331–339 (2004)
18. Iordache, R., Moldoveanu, F.: QoS-Aware Web Service Semantic Selection Based on Preferences. In: Int. Symposium on Intelligent Manufacturing and Automation. pp. 1152–1161 (2014)
19. Kaufer, F., Klusch, M.: WSMO-MX: A Logic Programming Based Hybrid Service Matchmaker. In: Proc. of the 4th European Conference on Web Services (ECOWS06). pp. 161–170. Zurich, Switzerland (2006)
20. Kifer, M., Lara, R., Polleres, A., Zhao, C., Keller, U., Lausen, H., Fensel, D.: A logical framework for web service discovery. In: Proceedings of the ISWC 2004 Workshop on Semantic Web Services (2004)
21. Klusch, M., Fries, B., Sycara, K.: OWLS-MX: a hybrid Semantic Web service matchmaker for OWL-S service. Journal of Web Semantics **7**(2), 121–133 (2009)
22. Klusch, M., Kapahnke, P.: Semantic Web Service Selection with SAWSDL-MX. In: Proc. of 2th Int. Workshop on Service Matchmaking and Resource Retrieval in the Semantic Web (SMRR08). pp. 3–18. Germany (2008)
23. Klusch, M., Nesbigall, S., Zinnikus, I.: MDSM: Model-Driven Semantic Web Service Matchmaking for Collaborative Business Processes. In: WI. pp. 612–618 (2008)
24. Stollberg, Michael, Keller, Uwe, Lausen, Holger, Heymans, Stijn: Two-Phase Web Service Discovery Based on Rich Functional Descriptions. In: Franconi, Enrico, Kifer, Michael, May, Wolfgang (eds.) ESWC 2007. LNCS, vol. 4519, pp. 99–113. Springer, Heidelberg (2007)
25. Lee, J., An Kao, H., Shanhu, Y.: Service Innovation and Smart Analytics for Industry 4.0 and Big Data Environment. In: Proc. of the 6th CIRP Conference on Industrial Product-Service Systems (2014)
26. Martin, D., Burstein, M., Hobbs, J., Lassila, O., McDermott, D., McIlraith, S., Narayanan, S., Paolucci, M., Parsia, B., Payne, T., Sirin, E., Srinivasan, N., Sycara, K.: OWL-S: Semantic Markup for Web Services, v1.1. Tech. rep., W3C (2004)
27. Ngan, L., Kanagasabai, R.: Semantic Web Service discovery: state-of-the-art and research directions. Personal and Ubiquitous Computing **17**, 1741–1752 (2013)
28. Plebani, P., Pernici, B.: URBE: Web Service Retrieval Based on Similarity Evaluation. TKDE **21**, 1629–1642 (2009)
29. Rodriguez-Garcia, M., Valencia-Garcia, R., Garcia-Sanchez, F., Samper-Zapater, J.: Ontology-based annotation and retrieval of services in the cloud. Knowledge-based Systems **56**, 15–25 (2014)
30. Staab, S., Studer, R., (eds.): Handbook on Ontologies. Springer (2009)

31. Stavropoulos, T., Gottis, K., Vrakas, D., Vlahavas, I.: aWESoME: a Web Service Middleware for Ambient Intelligence. Journal of Expert Systems With Applications **40**(11), 4380–4392 (2013)
32. Studer, R., Grimm, S., Abecker, A.: Semantic Web Services - Concepts, Technologies, and Applications. Springer (2007)
33. Vadivelou, G., Ilavarasan, E.: Performance evaluation of semantic approaches for automatic clustering of similar Web Services. In: IEEE World Congress on Computing and Communication Technologies. pp. 237–242 (2014)
34. Weise, T., Blake, M., Bleul, S.: Semantic Web Service Composition: The Web Service Challenge Perspective. In: Web Services Foundations. pp. 161–187 (2014)
35. Wu, J., Chen, L., Zheng, Z., Lyn, M., Wu, Z.: Clustering Web services to facilitate service discovery. Knowledge and Information Systems **38**, 207–229 (2014)
36. Zaremski, A., Wing, J.: Specification Matching of Software Components. ACM Transactions on Software Engineering and Methodology **6**(4), 333–369 (1997)

Maximal Set of XML Functional Dependencies for the Integration of Multiple Systems

Joshua Amavi[(✉)] and Mirian Halfeld Ferrari

INSA Centre Val de Loire, Univ. Orléans, LIFO EA 4022 FR-45067, Orléans, France
{joshua.amavi,mirian}@univ-orleans.fr

Abstract. A web application expected to deal with XML documents conceived on the basis of divers sets of (local) constraints would be expected to test documents with respect to all non contradictory constraints imposed by these original (local) sources. The goal of this paper is to introduce an optimized algorithm for computing the maximal set of XML functional dependencies (XFD) over multiple systems. The basis of our method is a sound and complete axiom system which is provided for relative XFD allowing two kinds of equality: value or node equality.

Keywords: XML · Functional dependencies · XFD · Interoperability

1 Introduction

This paper deals with the problem of exchanging XML (eXtensible Markup Language) data in a multi-system environment where a global central system should receive and process data coming from different local sources. Our global system is a conservative evolution of local ones. It conserves the possibility of accepting XML documents coming from any local (original) source. It extends local systems since it has its own schema and integrity constraints (generated from a merge of local ones) and may accept and deal with non-local XML documents (possibly non locally valid ones).

Our work aims at enriching schema evolution proposals by taking into account integrity constraints. Schema merging proposals are usually based on simple data models. Schemas can be more expressive than DTD and XSD, associated to integrity constraints (as in [6]) or expressed by a semantically richer data model (as in [23]).

A conservative schema evolution algorithm that extends minimally regular tree grammar is proposed in [9]. That approach for schema extension is inherently syntactic: only structural aspects of XML documents are considered and new grammars are built by syntactic manipulation of the original production rules. This paper aims at enriching that model by offering the possibility of computing from given local sets of XFD (XML Functional Dependencies), a cover of the biggest set of XFD that does not violate any local document. This is a first

Partially supported by CAPES-Stic/AmSud Project 052/2014 (SWANS).

© Springer-Verlag Berlin Heidelberg 2014
A. Hameurlain et al. (Eds.): TLDKS XIV, LNCS 8800, pp. 83–113, 2014.
DOI: 10.1007/978-3-662-45714-6_4

step towards an extension of a schema evolution proposal which will take into account integrity constraints. This extension intends to enrich schema evolution but is conceived as an independent procedure. In this way it may be applied or adapted to other schema evolution approaches.

Some applications of our work are: in the field of Digital Libraries, due to their need of evolution when new sources of data become available or when merging two libraries may be interesting [11]; in the construction of innovative services with data coming from diverse organizations that manipulate similar (though not identical) information, allowing us to envisage possible adaptations to big data applications [7]. In these cases, it is important to have a non contradictory set of integrity constraints (one that could be built from the original local constraints).

We suppose that $S_1, \ldots, S_n$ are local (original) systems which deal with sets of XML documents $X_1, \ldots, X_n$, respectively, and that inter-operate with a global, integrated system S. System S integrates local systems and is seen as an evolution of all of them. It can continue to receive information from any local (original) system, but it can also deal with information coming from other non local sources. Each set X_i conforms to schema constraints $\mathcal{D}_i$ and to integrity constraints $\mathcal{F}_i$ and follows an ontology O_i. Our goal is to associate system S to type and integrity constraints which represent a conservative evolution of local constraints. More precisely, given different triples $(\mathcal{D}_1, \mathcal{F}_1, O_1), \ldots, (\mathcal{D}_n, \mathcal{F}_n, O_n)$, we are interested in generating $(\mathcal{D}, cover\mathcal{F}, \mathcal{A})$, where:

(i) $\mathcal{D}$ is an extended type that accepts any local document;

(ii) $cover\mathcal{F}$ is a set of XFD equivalent to $\mathcal{F}$ the biggest set of functional dependencies (XFD), built from $\mathcal{F}_1, \ldots, \mathcal{F}_n$, that can be satisfied by all documents in $X_1, \ldots, X_n$ and

(iii) $\mathcal{A}$ is an ontology alignment that guides the construction of $\mathcal{D}$ and $\mathcal{F}$ in terms of semantics mapping. Notice that ontology issues are out of the scope of this paper, but we suppose the existence of $\mathcal{A}$ which is the basis of a pre-processing step where correspondence among tree paths (built on the different D_i) is established. The construction of this pre-processor is out of purpose in this paper; we just consider that the output of such pre-processing is an input of our algorithms.

This paper focus only on the generation of $cover\mathcal{F}$ which contains the XFD for which no violation is possible when considering document sets $X_1, \ldots, X_n$. It is important to notice that our algorithm is based on an axiom system and, thus, obtains $cover\mathcal{F}$ from $\mathcal{F}_1, \ldots, \mathcal{F}_n$, disregarding data.

The contribution of this paper is twofold. On one hand we introduce an axiom system together with the proofs of its soundness and completeness. On the other hand, we present an efficient way for computing, on the basis of our axiom system, the set $cover\mathcal{F}$. We prove that the obtained set $cover\mathcal{F}$ has good properties and some experiments show the efficiency of our approach.

The rest of this paper is organized as follows. Section 2 comments on some related work. Section 3 illustrates our goal with an example. Section 4 presents some background while Section 5 introduces our XFD. Section 6 focuses on our axiom system. Section 7 introduces our method for computing *cover$\mathcal{F}$* while in Section 8 we discuss on some experiments. Finally, Section 9 concludes the paper. We refer to [2] for the omitted proofs.

2   Related Work

Our motivation is to offer to a *global* system the capability of preserving the biggest set of local non contradictory constraints. Since the objective is to work only on constraint specification without any data involvement, we use an axiom system. To the best of our knowledge no other work considers this scenario.

We refer to [3, 13, 16, 19–21, 24] as other proposals for defining XFD and to [13, 22] for a comparison among some of them. Different XFD proposals entail different axiomatisation system, such as those in [13, 15, 21]. We adopt XFD presented in [4] for which we possess a validation tool (general algorithm in [6]). The approach in [12] defines XFD as tree queries, which implies a complex implementation, and proposes static XFD validation *w.r.t.* updates.

To achieve our goal out first task is to propose an axiom system for the adopted XFD, together with an efficient algorithm for computing the closure of a set of paths. Our work on this axiom system is comparable to the one proposed in [21]. The main differences are: (i) we propose a more powerful path language allowing the use of a wild-card; (ii) our XFD are verified *w.r.t.* a context and not only *w.r.t.* the root, *i.e.*, XFD can be relative; (iii) our XFD can be defined by taking into account two types of equality: value and node equality and (iv) we use simpler concepts (such as branching paths, projection) which, we believe, allow us to prove that our axiom system is sound and complete in a clearer way.

We use our axiom system in the development of a practical tool: to filter local XFD in order to obtain a set containing only XFD that cannot be violated by any local XML document. Our global system aims to deal with data coming from any local source, but not to perform data fusion. Thus, our work presents an original point of view, since we are not interested in putting together all the local information, but just in manipulating them. Usually, schema integration proposal comes together with the idea of data fusion. XML data fusion is considered in papers such as [8, 18]. Data exchange is considered in [10] that aims to construct an instance over a target schema, based on the source and a given mapping, and to answer queries against the target data, consistently with the source data.

Proposals concerning XML type evolution usually do not take into account the evolution of associated integrity constraints which are extremely important in the maintenance of consistent information. In [14] authors offer as a perspective to apply to XML their proposal of adapting functional dependencies according to schema changes. This is done in [19] where authors consider the problem of constraint evolution in conformance with type evolution. The type evolution in [9] is well adapted to our purposes; it seems possible to combine it with our XFD filter in order to generate a set of constraints allowing interoperability.

3   Motivating Example

We suppose universities or educational institutions, from the same region in France, which want to implement a central data system to obtain and process information concerning their courses and students, independently of local systems already in use. Their goal is to obtain a central system that ensures a maximum number of the local non-contradictory integrity constraints.

Each educational institution has established, locally, its own constraints. For instance, let the XML trees in Figure 1 be documents from two different universities. Each document is valid *w.r.t.* the functional dependencies presented in Table 1, *i.e.*, documents in X_1 are valid *w.r.t.* $\mathcal{F}_1$, those in X_2 are valid *w.r.t.* $\mathcal{F}_2$. We recall that local schemas and concepts may be different.

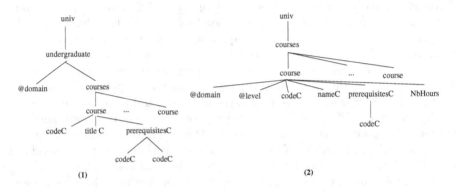

Fig. 1. Two XML documents from different local sources

In the XML domain, a functional dependency (XFD) is defined by paths over a tree. Each path selects a node on a tree. Values or positions of the selected nodes are gathered to build tuples that will be used to verify whether a given XML document satisfies an XFD. For example, consider the XFD *f: (univ, (undergraduate/courses/course/codeC → undergraduate/courses/course/titleC))* on the first document of Figure 1(1). It specifies that the context is *univ*, *i.e.*, that the constraint should be verified on data below a node labelled *univ*. In this context, *f* entails the construction of tuples composed by values obtained by following the paths: *univ/undergraduate/courses/course/codeC, univ/undergraduate/-courses/course/titleC.* As in the relational model, a document is valid *w.r.t. f* if any two tuples agreeing on values obtained from *univ/undergraduate/courses/course/codeC* also agree on values obtained from *univ/undergraduate/courses/course/titleC.* Thus, in a university the code of a course determines its name.

Similarly, the XFD *f1: (univ, (undergraduate/courses/course/codeC → undergraduate/courses/course/prerequisitesC))* entails tuples where the path *univ/undergraduate/courses/course/prerequisitesC* leads us to obtain sub-trees having roots labelled *prerequisitesC* (*i.e.*, sub-trees containing information about prerequisites). This constraint indicates that courses having the same code should have the same prerequisites. A document is valid *w.r.t.* $f1$ if any two tuples agreeing on values obtained from *univ/undergraduate/courses/course/codeC* also agree on values obtained from *univ/undergraduate/courses/course/prerequisitesC*, *i.e.*, obtained sub-trees are isomorphic.

Table 1. XFD in $\mathcal{F}_1$ and $\mathcal{F}_2$

$\mathcal{F}$	XFD
1	$(univ, (undergraduate/courses/course/codeC \rightarrow undergraduate/courses/course/titleC))$
1	$(univ, (undergraduate/courses/course/codeC \rightarrow undergraduate/courses/course/prerequisitesC))$
1	$(univ, (undergraduate/courses/course/codeC \rightarrow undergraduate/@domain))$
2	$(univ, (courses/course/codeC \rightarrow courses/course/nameC))$
2	$(univ, (courses/course/codeC \rightarrow courses/course/@domain))$
2	$(univ, (courses/course/codeC \rightarrow courses/course/@level))$
2	$(univ, (\{courses/course/nameC, courses/course/@level\} \rightarrow courses/course/NbHours))$

Now consider the first three XFD in Table 1, concerning source 1. They indicate that in a university, the code of a course determines its name, its domain and its prerequisites. In other words, a course is identified by its code.

From the alignment of local ontologies we assume that Table 2 is available, making the correspondence among paths on the different local sets of documents. Thus, it is possible to conclude that, for instance, XFD *f: (univ, (undergraduate/courses/course/codeC → undergraduate/courses/course/titleC))* and *(univ, (courses/course/codeC → courses/course/nameC))* are equivalent, *i.e.*, they represent the same constraint since they involve the same concepts: in a university, the code of a course determines its name.

Table 2. Extract of the translation table

Paths from $\mathcal{D}_1$	Paths from $\mathcal{D}_2$
univ/undergraduate/courses/course/codeC	*univ/courses/course/codeC*
univ/undergraduate/courses/course/titleC	*univ/courses/course/nameC*
univ/undergraduate/courses/course/prerequisitesC	*univ/courses/course/prerequisitesC*
univ/undergraduate/courses/course/prerequisitesC/codeC	*univ/courses/course/prerequisitesC/codeC*
univ/undergraduate/@domain	*univ/courses/course/@domain*

Assuming that we have only these two local sources, we want to obtain, from $\mathcal{F}_1$ and $\mathcal{F}_2$, the biggest set of XFD $\mathcal{F}$ that does not contradict any document in X_1 and X_2. To reach this goal, we should consider *all* XFD derivable from $\mathcal{F}_1$ and $\mathcal{F}_2$, which may result in very big sets of XFD. Indeed, the set $\mathcal{F}$ is, usually, a very big one - too big to work with. A better solution consists of computing *cover*$\mathcal{F}$, *a cover of* $\mathcal{F}$ (*i.e.*, a (usually) smaller set of XFD that is equivalent to $\mathcal{F}$), without computing *all* XFD derivable from $\mathcal{F}_1$ and $\mathcal{F}_2$. In this paper, we propose an algorithm that generates this set of XFD.

In our example, the resulting *cover*$\mathcal{F}$ would contain XFD of Table 3. Let us analyse this solution. In Table 3, the first and the fourth XFD are equivalent. They are kept in *cover*$\mathcal{F}$ since all documents in X_1 and X_2 are valid *w.r.t.* it. The same reasoning is applied for the second and third XFD in Table 3. The two last XFD involve concepts that occur only in X_2 and, thus, cannot be violated by documents

Table 3. XFD in the resulting $\mathcal{F}$

1 $(univ, (undergraduate/courses/course/codeC \rightarrow undergraduate/courses/course/titleC))$
2 $(univ, (undergraduate/courses/course/codeC \rightarrow undergraduate/@domain))$
3 $(univ, (courses/course/codeC \rightarrow courses/course/@domain))$
4 $(univ, (courses/course/codeC \rightarrow courses/course/nameC))$
5 $(univ, (courses/course/codeC \rightarrow courses/course/@level))$
6 $(univ, (\{courses/course/nameC, courses/course/@level\} \rightarrow courses/course/NbHours))$

in X_1. Notice that the XFD $(univ, (undergraduate/courses/course/codeC \rightarrow undergraduate/courses/course/prerequisitesC))$ in $\mathcal{F}_1$, which states that courses with the same code have the same set of prerequisites, is not in $\mathcal{F}$. The reason is that according to the ontology alignment, this XFD is equivalent to $(univ, (courses/course/codeC \rightarrow courses/course/prerequisitesC))$ in $\mathcal{F}_2$. However, as $\mathcal{F}_2$ does not contain this XFD, documents in X_2 may violate it (since the involved concepts exist in X_2).

4   Preliminaries

Our work uses XFD such as those in [4,6]. An XML document is seen as a tuple $\mathcal{T} = (t, type, value)$. The tree t is the function t: $dom(t) \rightarrow \Sigma$ where: (A) $\Sigma = \Sigma_{ele} \cup \Sigma_{att} \cup \{data\}$ is an alphabet; Σ_{ele} is the set of element names and Σ_{att} is the set of attribute names and (B) $dom(t)$ is the set of positions numbered according to Dewey encoding. Given a tree position p, function $type(t, p)$ returns a value in $\{data, element, attribute\}$. Similarly, $value(t, p) =$
$$\begin{cases} p & \text{if } type(t, p) = element \\ val \in \mathbf{V} & \text{otherwise} \end{cases}$$
where $\mathbf{V}$ is an infinite recursively enumerable domain.   □

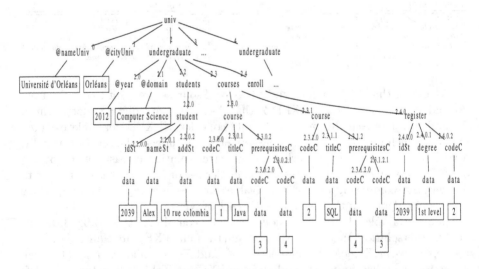

Fig. 2. XML document concerning the first degree (undergraduate) at a university

As many other authors, we distinguish two kinds of equality in an XML tree, namely, *value equality and node equality*. Two nodes are *value equal* when they are roots of isomorphic sub-trees. Two nodes are *node equal* when they have the same position number. To combine both equality notions we use the symbol E, that can be represented by V for value equality, or N for node equality. Our value equality definition does not take into account the document order. For instance, in Figure 2, nodes in positions 2.3.0.2 and 2.3.1.2 are value equal, but nodes 2.3.0 and 2.3.1 are not.

4.1 Linear Paths

Linear paths are used to address parts of an XML document. Let *PL* **be the language where a path is defined** by $\rho ::= [] \mid l \mid \rho/\rho \mid \rho//l$ where $[]$ is the empty path, l is a label in Σ, ”/” is the concatenation operation, ”//” represents a finite sequence (possibly empty) of labels. Notice that $l/[] = []/l = l$ and $[]//l = //l$. We distinguish between paths using the wild-card $//$ and *simple paths* (those with no wild-card) and we denote by $\mathbb{P}$ the set of *all possible rooted simple paths that may occur in an XML tree t respecting a given schema $\mathcal{D}$*.

In this work we consider that the set $\mathbb{P}$ is generated from a given schema $\mathcal{D}$. Notice that $\mathbb{P}$ is a *finite* set of simple (top-down) paths and, in this way, the schema from which it is obtained should ensure a limited *depth* of label repetitions. In other words, the language $L(\mathcal{D})$, obtained from a finite state automaton $\mathcal{D}$ (which is built from a given type $\mathcal{D}$), should be finite. Such kind of schema can be expressed, for instance, by a non-recursive DTD. In this way, we are more general than [4,6], where $\mathbb{P}$ was the set containing only all possible paths in *one* given tree.

It is important to notice that one path with wild-card can be associated to a *set* of simple paths in $\mathbb{P}$. This set of simple paths is the language $L(A_P)$, where A_P is a finite-state automaton (FSA) obtained on the basis of the two following steps:

1. From the path language P we construct a finite-state automaton B_P which recognizes the expression P in PL and is similar to restricted regular expressions.
2. $A_P = D \cap B_P$. We retain in $L(A_P)$ the paths which respect the path language P and are simple paths in $\mathbb{P}$.

Example 1. We suppose a DTD concerning undergraduate course in a university such that the document of Figure 2 is valid *w.r.t.* it. Let D be the FSA that recognizes $\mathbb{P}$, the language of prefixes of the paths defined by this DTD. Let $P = univ/undergraduate//codeC$ and B_P the FSA that recognizes P. The set $L(B_P) \cap \mathbb{P}$ contains only the simple paths in $L(B_P)$ that trees respecting $\mathbb{P}$ may have, *i.e.*, { *univ/undergraduate/courses/course/codeC, univ/undergraduate/courses/course/prerequisitesC/codeC, univ/undergraduate/enroll/register/codeC*}. □

A path P is **valid** if: (*i*) it conforms to the syntax of PL, (*ii*) $L(A_P) \neq \emptyset$, (*iii*) for all label $l \in P$, if $l = data$ or $l \in \Sigma_{att}$, then l is the last symbol in P.

In this work, given a path P in PL we define the following functions:

- $Last(P) = l_n$ where l_n is the last label on path P.
- $Parent(P) = \{l_1/\ldots/l_{n-1} \mid l_1/\ldots/l_{n-1}/l_n \in L(A_P)$ for $n > 1\}$, the set of simple paths starting at a node labelled by l_1 and ending at the parent of l_n (where $Last(P) = l_n$).
- A path Q is a prefix of P (we note $Q \preceq_{PL} P$) if $L(A_Q) \subseteq L(PREFIX(A_P))$ where $PREFIX(A_P)$ is the finite state automaton that accepts the language containing all prefixes of $L(A_P)$.
- The **longest common prefix** (or the intersection) of P and Q, denoted by $P \cap Q$, describes the set of simple paths $\{P' \cap Q' \mid P' \in L(A_P) \wedge Q' \in L(A_Q)\}$. The longest common prefix of two simple paths P' and Q' (denoted $P' \cap Q'$) is the simple path R where $R \preceq P'$ and $R \preceq Q'$ and there is no path R' such that $R \prec R'$, $R' \preceq P'$ and $R' \preceq Q'$.

Example 2. Consider the XML document of Figure 2. The simple path *univ/undergraduate/courses* is a prefix for *univ/undergraduate/courses/course/codeC*. Given $P' =$*univ/undergraduate/courses/course/codeC* and $Q' =$ *univ/undergraduate/courses/course/prerequisitesC/codeC*, their longest common prefix is *univ/undergraduate/courses/course*.
Given $P =$ *univ//codeC* and $Q =$ *univ//idSt*, the longest common prefix $P \cap Q =$ { *univ/undergraduate, univ/undergraduate/enroll/register*}   □

Now, let $I = p_1/\ldots/p_n$ be a sequence of positions such that each p_i is a direct descendant of p_{i-1} in t. Then I is an **instance of a path** P **over a given tree** t if and only if the sequence $t(p_1)/\ldots/t(p_n) \in L(A_P)$. We denote by $Instances(P,t)$ the set of all instances of P over t. Functions $Last$, $Parent$, $Prefix$ and the *longest common prefix* are extended to path instances in the obvious manner. Notice that the longest common prefix allows the identification of the least common ancestor.

We now remark that, in this paper, we will only deal with **complete trees** (*i.e.*, documents with no missing information). Let $\mathbb{P}$ be a set of simple paths associated to an XML document $\mathcal{T}$. We say that $\mathcal{T}$ is complete w.r.t. $\mathbb{P}$ if whenever there exists paths P and P' in the associated $\mathbb{P}$ such that $P' \prec P$ and there exist an instance I' for P' such that node v' is the last node in I', then there exists an instance I for P such that v is the last node in I and v' is an ancestor of v. For example, let $\mathbb{P}$ be a set containing paths $R/A/C$, $R/A/D$, R/B and their prefixes. Then, representing trees as terms, we notice that $R(A(C,D),B)$, $R(A(C,D),A(C,D),B)$ are complete trees, while $R(A(C),B)$, $R(A(C,D))$ are no complete trees.

Given two valid paths P and Q over a tree t, we want to verify whether two given path instances match on the longest common prefix of P and Q. To this end we define **the boolean function** $isInst\_lcp(P,I,Q,J)$ which returns *true* when all the following conditions hold: (*i*) $I \in Instances(P,t)$; (*ii*) $J \in Instances(Q,t)$ and (*iii*) $I \cap J$ is an instance of a path in $P \cap Q$; otherwise, it returns *false*.

4.2   Branching Paths

Now we introduce the notion of branching paths also called a pattern in the literature [5,20]. A branching path is a non-empty set of simple paths having a common prefix. The projection of a tree over a branching path determines the tree positions corresponding to the given path. Thus, as defined below, this projection is a set of prefix closed simple path instances that respect some important conditions.

Definition 1 (Branching path). *A branching path is a finite set of prefix-closed (simple) paths on a tree t.*   □

Definition 2 (Projection of a tree T over a branching path M). *Let M be a branching path over a tree T. Let $Long_M$ be the set of paths in M that are not prefix of other paths in M. Let SetPathInst be the set of (simple) path instances that verifies:*

1. *For all paths $P \in Long_M$ there is one and only one instance inst $\in$ Instances(P,t) in the set SetPathInst.*
2. *For all inst $\in$ SetPathInst there is a path $P \in Long_M$ such that inst $\in$ Instances(P,t).*
3. *For all instances inst and inst' in SetPathInst, if inst $\in$ Instances(P,t) and inst' $\in$ Instances(Q,t), then isInst_lcp$(P,inst,Q,inst')$ is true.*

A projection of T over M, denoted by $\Pi_M(T)$, is a tuple $(t^i, type^i, value^i)$ where $type^i(t^i,p) = type(t,p)$, $value^i(t^i,p) = value(t,p)$ and t^i is a function $\Delta \to \Sigma$ in which:

- $\Delta = \bigcup_{inst \in SetPathInst} \{p \mid p \text{ is a position in inst}\}$
- $t^i(p) = t(p)$, $\forall p \in \Delta$ □

Given the projection of two branching paths, $\Pi_{M_1}(T)$ and $\Pi_{M_2}(T)$, the union $\Pi_{M_1}(T) \cup \Pi_{M_2}(T)$ is naturally obtained by considering all the path instances used to obtain each projection.

Example 3. Consider the XML document of Figure 2. Let M be a branching path defined from the set {*univ/undergraduate/courses/course/codeC, univ/undergraduate/courses/course/prerequisitesC*}, *i.e.*, M contains these paths and all their prefixes. An example of a projection of T over M is the one where $t(\epsilon) = univ$, $t(2) = undergraduate$, $t(2.3) = courses$, $t(2.3.0) = course$, $t(2.3.0.0) = codeC$ and $t(2.3.0.2) = preresquisitesC$. However, if we take $t(\epsilon) = univ$, $t(2) = undergraduate$, $t(2.3) = courses$, $t(2.3.0) = course$, $t(2.3.0.0) = codeC$ and $t(2.3.1.2) = preresquisitesC$, we do not have a projection of T over M. Indeed, in Definition 2, if we consider $P = univ/undergraduate/courses/course/codeC$ and its instance $inst = \epsilon/2/2.3/2.3.0/2.3.0.2$ together with $Q = univ/undergraduate/courses/course/prerequisitesC$ and its instance $inst' = \epsilon/2/2.3/2.3.1/2.3.1.2$ we obtain isInst_lcp$(P,inst,Q,inst') = false$. Notice that the longest common paths $P \cap Q$ is *univ/undergraduate/courses/course*.   □

From Definition 2, we remark that the projection of T over a branching path M contains exactly one instance of every path in M. In the following, when needed, *we denote by* $\Pi_M(T)[P]$ *the unique instance of the simple path P in* $\Pi_M(T)$. Indeed, when we write $\Pi_M(T)[P]$ we restrict the projection of T over M to the instance (in the projection) of one simple path P.

Lemma 1. *Let* $\Pi_M(T)$ *be a projection of a tree T over a branching path M. For each two simple paths P and Q in M if $I = \Pi_M(T)[P]$ and $J = \Pi_M(T)[Q]$ then we have isInst_lcp$(P, I, Q, J) = true$.*   □

Now we are interested in building a relation where each tuple corresponds to values determined by a given projection $\Pi_M(T)$.

Definition 3 (Tuple obtained from Projection). *Let M be a branching path and $X = \{P_1, \ldots, P_k\}$ be a set of paths such that $X \subseteq M$. Let $\tau = \Pi_M(T) = (t^i, type^i, value^i)$ be a projection of the tree T on M. Let $I_j = \Pi_M(T)[P_j]$ be the only instance of path P_j in $\Pi_M(T)$ where $j \in [1, \ldots, k]$. The tuple corresponding to X on τ, denoted by $\tau[X]$, is defined as[1]*
$$\tau[X] = (P_1 : value^i(t^i, Last(I_1)), \ldots, P_k : value^i(t^i, Last(I_k))).$$
We denote by $\tau[P_j]$ the result of $P_j : value^i(t^i, Last(I_j))$. Two tuples $\tau^1[X]$ and $\tau^2[X]$ are equal w.r.t. the equality list $E = (E_1, \ldots, E_k)$, denoted by $\tau^1[X] =_E \tau^2[X]$, iff $\forall j \in [1 \ldots k]$, $\tau^1[P_j] =_{E_j} \tau^2[P_j]$.   □

The tuple $\tau[X]$ is formed by the values or nodes found in an XML document T from a projection on branching path M, and is constructed by following the named perspective in relational database [1] where the name of attributes in the tuples are known. Notice also that the equality between two tuples may involve different kinds of equality, one for each path.

5   Functional Dependencies in XML

Usually, a functional dependency in XML (XFD) is denoted by $X \rightarrow Y$ (where X and Y are sets of paths) and it imposes that for each pair of tuples (Definition 3) t_1 and t_2 if $t_1[X] = t_2[X]$ then $t_1[Y] = t_2[Y]$. In this paper, our XFD are defined as those in [5,6], generalizing the proposals in [3,17,20,22]. As the dependency can be imposed in a specific part of the document, we specify a *context path*.

Definition 4 (XML Functional Dependency). *Given an XML tree t, an XFD f is an expression of the form:*
$$f = (C, (\{P_1 \ [E_1], \ldots, P_k \ [E_k]\} \rightarrow \{Q_1 \ [E_1'], \ldots, Q_m \ [E_m']\}))$$
where C is a path that starts from the root of t (context path) ending at the context node; $\{P_1, \ldots, P_k\}, \{Q_1, \ldots, Q_m\}$ are non-empty sets of paths in t. Both P_i ($i \in [1, \ldots, k]$) and Q_i ($i \in [1, \ldots, m]$) start at the context node. The set

---
[1] If it is clear by the context, we omit the path when showing a tuple.

$\{P_1, \ldots, P_k\}$ *is the left-hand side (LHS) or determinant of an XFD, and the set* $\{Q_1, \ldots, Q_m\}$ *is the right-hand side (RHS) or the dependent paths. The symbols* $E_1, \ldots, E_k, E'_1, \ldots, E'_m$ *represent the equality type associated to each dependency path. When symbols* $E_1, \ldots, E_k$ *or* $E'_1, \ldots, E'_m$ *are omitted, value equality is the default choice.* □

Notice that in an XFD the set of paths $\{C/P_1, \ldots, C/P_k, C/Q_1, \ldots, C/Q_m\}$ defines branching paths and that, as in [22], our XFD definition allows the combination of two kinds of equality.

Definition 5 (XFD Satisfaction). *Let* $\mathcal{T}$ *be an XML document and* $f = (C, (\{P_1 [E_1], \ldots, P_k [E_k]\} \rightarrow \{Q_1 [E'_1], \ldots, Q_m [E'_m]\}))$ *an XFD. Let* M *be a branching path defined from* f. *We say that* $\mathcal{T}$ *satisfies* f *(noted by* $\mathcal{T} \models f$) *if and only if for all* $\tau^1 = \Pi_M^1(\mathcal{T})$ *and* $\tau^2 = \Pi_M^2(\mathcal{T})$ *that are projections of* $\mathcal{T}$ *on* M *and that coincide at least on their prefix* C, *we have:* *If* $\tau^1[C/P_1, \ldots, C/P_k] =_E \tau^2[C/P_1, \ldots, C/P_k]$ *then* $\tau^1[C/Q_1, \ldots, C/Q_m] =_{E'} \tau^2[C/Q_1, \ldots, C/Q_m]$ *where* $E = (E_1, \ldots, E_k)$ *and* $E' = (E'_1, \ldots, E'_m)$. □

Example 4. Consider the following XFD on the document of Figure 2.

XFD1: $univ//courses, (\{course/codeC\} \rightarrow course/titleC)$
Considering the set of courses of an undergraduate domain, courses having the same code have the same title.
XFD2: $univ, (\{undergraduate//course/codeC\} \rightarrow undergraduate//course/titleC)$. Considering the set of all courses in a university, courses having the same code have the same title.
XFD3: $univ//students, (\{student/idSt\} \rightarrow student[N])$. Considering the set of students of an undergraduate domain, no two students have the same number and each student appears once. □

An XML document $\mathcal{T}$ satisfies a set of XFD $\mathcal{F}$, denoted by $\mathcal{T} \models \mathcal{F}$, if $\mathcal{T} \models f$ for all f in $\mathcal{F}$. Usually it is important to reason whether a given XFD f is also satisfied on $\mathcal{T}$ when $\mathcal{F}$ is satisfied. The following definition introduces this notion.

Definition 6 (XFD Implication). *Given a set* $\mathcal{F}$ *of XFD we say that* $\mathcal{F}$ *implies* f, *denoted by* $\mathcal{F} \models f$, *if for every XML tree* $\mathcal{T}$ *such that* $\mathcal{T} \models \mathcal{F}$ *then* $\mathcal{T} \models f$. □

Based on the notion of implication we can introduce the definition of closure for a set of XFD.

Definition 7 (Closure of a set of XFD). *The closure of a set of XFD* $\mathcal{F}$, *denoted by* $\mathcal{F}^+$, *is the set containing all the XFD which are logically implied by* $\mathcal{F}$, *i.e.,* $\mathcal{F}^+ = \{f \mid \mathcal{F} \models f\}$. □

Notation: In the rest of this paper, given an XFD $(C, (X \rightarrow A))$ where $X = \{P_1, \ldots, P_n\}$ is a set of paths and A is a path, we use C/X as a shorthand for the set $\{C/P_1, \ldots, C/P_n\}$.

6 Axiom System

To find which XFD f are also satisfied when a given set of XFD $\mathcal{F}$ is satisfied we need inference rules that tell how one or more dependencies imply other XFD. In this section we present our axiom system and prove that it is sound (we cannot deduce from F any false XFD) and complete (from a given set $\mathcal{F}$, the rules allow us to deduce all the true dependencies). Our axiom system is close to the one proposed in [21], but has two important differences: our XFD are defined *w.r.t.* a context (and not always *w.r.t.* the root) and we use two kinds of equality.

Definition 8 (Inference Rules for XFD). *Given a tree T and XFD defined over paths in $\mathbb{P}$, our axioms are:*

A1: **Reflexivity** $(C, (\{P_1\,[E_1], \ldots, P_n\,[E_n]\} \to P_i\,[E_i]))$, $\forall i \in [1 \ldots n]$.

A2: **Augmentation** *If $(C, (\{P_1\,[E_1], \ldots, P_n\,[E_n]\} \to \{Q_1\,[E_1'], \ldots, Q_m\,[E_m']\}))$ then $(C, (\{R\,[E_r], P_1\,[E_1], \ldots, P_n\,[E_n]\} \to \{R\,[E_r], Q_1\,[E_1'], \ldots, Q_m\,[E_m']\}))$.*

A3: **Transitivity** *If $(C, (\{P_1\,[E_1], \ldots, P_n\,[E_n]\} \to \{Q_1\,[E_1'], \ldots, Q_m\,[E_m']\}))$ and $(C, (\{Q_1\,[E_1'], \ldots, Q_m\,[E_m']\} \to S\,[E_s]))$ then $(C, (\{P_1\,[E_1], \ldots, P_n\,[E_n]\} \to S\,[E_s]))$.*

A4: **Branch Prefixing** *If $(C, (\{P_1'\,[E_1'], \ldots, P_n'\,[E_n']\} \to P_{n+1}\,[E_{n+1}]))$ and there exist paths $C/P_1, \ldots, C/P_n$ (not necessarily distinct) such that*
(i) $P_i' \cap P_{n+1} \preceq_{PL} P_i$ and
(ii) $P_i \preceq_{PL} P_i'$ or $P_i \preceq_{PL} P_{n+1}$
then $(C, (\{P_1\,[E_1], \ldots, P_n\,[E_n]\} \to P_{n+1}\,[E_{n+1}]))$.

A5: **Ascendency** *If Q is a prefix for P then $(C, (P\,[N] \to Q\,[N]))$.*

A6: **Attribute Uniqueness** *If $Last(P) \in \Sigma_{att}$ then $(C, (Parent(P)\,[E] \to P\,[E]))$.*

A7: **Root Uniqueness** $(C, (\{P_1\,[E_1], \ldots, P_n\,[E_n]\} \to []\,[E_{n+1}]))$.

A8: **Context Path Extension** *If $(C, (\{P_1\,[E_1], \ldots, P_n\,[E_n]\} \to P_{n+1}\,[E_{n+1}]))$ and there is a path Q such that $P_1 = Q/P_1', \ldots, P_{n+1} = Q/P_{n+1}'$ then $(C/Q, (\{P_1'\,[E_1], \ldots, P_n'\,[E_n]\} \to P_{n+1}'\,[E_{n+1}]))$.*

A9: **Node Equality to Value Equality** $\forall P$, $(C, (P\,[N] \to P\,[V]))$.

Example 5. A given university has one or more undergraduate specialities (first degree) and, for each of them, we store its domain and year together with information concerning students, courses and enrolment. Figure 2 shows a part of this XML document over which we illustrate the intuitive meaning of axioms A4-A9. The intuition of the three first axioms (A1-A3) is the same as in relational.

A4: If $(univ, (\{undergraduate/@domain, courses//codeC\} \to undergraduate/enroll//degree))$ then we can say that: $(univ, (\{undergraduate/@domain, undergraduate/courses\} \to undergraduate/enroll//degree))$ or $(univ, (\{undergraduate/@domain, undergraduate/courses/course\} \to undergraduate/enroll//degree))$ or $(univ, (undergraduate \to undergraduate/enroll//degree))$.

The initial XFD states that all courses having $codeC$ in the same domain correspond to the same degree. From this XFD, we can deduce, among others, the XFD $(univ, (undergraduate \rightarrow undergraduate/enroll//degree))$ stating that an undergraduate speciality is associated to only one degree ($e.g.$, Bachelor's).

A5: Given a path $P = undergraduate//register/idSt$, we can derive $(univ, (\{undergraduate//register/idSt\} \rightarrow undergraduate//register))$.

A6: Given $P = undergraduate/@year$, we derive that $(univ, (\{undergraduate[N]\} \rightarrow undergraduate/ @year[N]))$.

A8: If $(univ/undergraduate, (students/student/idSt \rightarrow students/student/nameSt))$ then $(univ/undergraduate/students, (student/idSt \rightarrow student/nameSt))$. If, in the context of an undergraduate domain, the $idSt$ identifies the name of a student; this is also true in the context of $students$.

A9: When we have a node equality, for instance, for $univ/undergraduate//course$, it means that we are considering a specific, uniquely referred, course in our document. Thus, $(univ, (\{undergraduate//course[N]\} \rightarrow undergraduate//course[V]))$ is a valid XFD.

Notice that $A5$ does not hold when dealing with value equality. The tree on Figure 2 violates the XFD $(univ, (\{undergraduate//course/prerequisitesC[V]\} \rightarrow undergraduate//course[V]))$. Indeed $Last(2.3.0.2) =_V Last(2.3.1.2)$ but $Last(2.3.0) \neq_V Last(2.3.1)$.

Remark that although we have value equality, the following rule $(C, (P[V] \rightarrow P/Q[V]))$ does not hold. Let us consider the XFD $(univ, (\{undergraduate//prerequisitesC[V]\} \rightarrow undergraduate//prerequisitesC/codeC[V]))$. The tree on Figure 2 does not satisfy this XFD because we have $Last(2.3.0.2) =_V Last(2.3.1.2)$ but $Last(2.3.0.2.0) \neq_V Last(2.3.1.2.0)$.   □

The set of axioms in Definition 8 establishes an inference system with which one can derive other XFD.

Definition 9 (XFD Derivation). *Given a set $\mathcal{F}$ of XFD, we say that an XFD f is derivable from the functional dependencies in $\mathcal{F}$ by the set of inference rules in Definition 8, denoted by $\mathcal{F} \vdash f$, if and only if there is a sequence of XFD $f_1, f_2, \ldots, f_n$ such that (i) $f = f_n$ and (ii) for all $i = 1, \ldots, n$ the XFD f_i is in $\mathcal{F}$ or it is obtainable from $f_1, f_2, \ldots f_{i-1}$ by means of applying an axiom A1-A9 (from Definition 8).*   □

Our axiom system is sound and complete. The proofs are summarized in Appendix A and B and in [2] one can find more detailed versions. Additional inferences rules (Union, Decomposition, Pseudotransitivity and Subtree Uniqueness) can be derived from axioms of Definition 8 as we show in [2]. Notice that as we have the Union and Decomposition axioms, an important consequence is that an XFD $(C, (\{P_1 [E_1], \ldots, P_k [E_k]\} \rightarrow \{Q_1 [E'_1], \ldots, Q_m [E'_m]\}))$ holds if and only if $(C, (\{P_1 [E_1], \ldots, P_k [E_k]\} \rightarrow \{Q_i [E'_i]\}))$ holds for $i \in [1, \ldots, m]$. Thus, having a single path on the right-hand side of an XFD is sufficient. Once we have our axiom system, we can define the closure of a set of paths $w.r.t.$ a set of XFD.

Definition 10 (Closure of a set of Paths). *Let X be a set of paths and let C be a path defining a context. Let $E = (E_1, \ldots, E_n)$ be the equality list associated to X. The closure of (C, X) with respect to $\mathcal{F}$, denoted by $(C, X[E])^+_{\mathcal{F}}$, is the set of paths $\{C/P_1[E'_1], \ldots, C/P_m[E'_m]\}$ such that $(C, (X[E] \rightarrow \{P_1[E'_1], \ldots, P_m[E'_m]\}))$ can be deduced from $\mathcal{F}$ by the axiom system in Definition 8. In other words, $(C, X[E])^+_{\mathcal{F}} = \{C/P[E'] \mid \mathcal{F} \vdash (C, (X[E] \rightarrow P[E']))\}$. When there is no ambiguity about the set $\mathcal{F}$ being used, we just note $(C, X[E])^+$.* □

To compute $(C, X[E])^+$ we start with a set T containing all the prefixes of the paths in X. Then we build a set V containing all the paths ending on attributes and having a path in T as its parent. Starting with $X^{(0)} = T \cup V$ we compute each $X^{(i+1)}$ from $X^{(i)}$ by applying the axiom system on $\mathcal{F}$. At each step, new sets T and V are computed and added to $X^{(i)}$. The loop ends when no new path can be added to $X^{(i)}$. In [2] we present this algorithm together with the proof of its soundness and completeness.

We also define two other functions, namely *closure1Step* and *inverseClosure1Step*. Function *closure1Step* computes one step of the closure of a set of paths. Its implementation consists in applying the same algorithm used to find $(C, X[E])^+$, in order to compute $X^{(1)}$. The result is a set of paths. Function *inverseClosure1Step* considers XFD inversely and computes an "inverse closure" one step backward. For instance, given a set of paths X the function finds all sets of paths Z for which we have $C/Z \rightarrow C/X$. The computed result is a set S containing sets of paths.

These functions are going to be used in the following section in order to compute *cover$\mathcal{F}$*.

7   Computing Functional Dependencies for Interoperability

7.1   Algorithm for Computing *cover$\mathcal{F}$*

Given XFD sets $\mathcal{F}_1, \ldots, \mathcal{F}_n$, let $\mathcal{F} = (\mathcal{F}_1^+ \cap \cdots \cap \mathcal{F}_n^+) \cup (K_1 \cup \cdots \cup K_n)$ where for $1 \leq i \leq n$, $\mathcal{F}_i^+$ is the closure of $\mathcal{F}_i$ and K_i is a set of XFD containing all XFD f which can be obtained from $\mathcal{F}_i$ but that cannot be violated by documents in X_j (for $j \neq i$). In [2] we have proved that $\mathcal{F}$ is the biggest set of XFD that are not in contradiction with any set $\mathcal{F}_1, \ldots, \mathcal{F}_n$. In other words, all documents in $X_1, \ldots, X_n$ valid *w.r.t.* $\mathcal{F}_1, \ldots, \mathcal{F}_n$ should stay valid *w.r.t.* $\mathcal{F}$. Our goal is to propose an algorithm that computes a set *cover$\mathcal{F}$* which is equivalent to $\mathcal{F}$ (*cover$\mathcal{F} \equiv \mathcal{F}$*), and usually the number of XFD in *cover$\mathcal{F}$* is much smaller than the number of XFD in $\mathcal{F}$.

Algorithm 1 generates *cover$\mathcal{F}$* as expected. As input, the algorithm receives the local sets of XFD together with the set of possible paths given by each local schema. Notice that for the sake of simplicity, we suppose only two local sources, but Algorithm 1 can be easily extended for n local sources. Translation functions Φ_1 and Φ_2 are available. These functions work on the translation table (obtained from the ontology alignment $\mathcal{A}$): given a path P from, for instance $\mathbb{P}_2$, $\Phi_1(P)$ gives its equivalent path in $\mathbb{P}_1$, if it exists; otherwise it returns the identity. The function Φ_2 works on a symmetric way. Indeed, we note i and $\bar{i}$ to indicate symmetric sources (*e.g.*, when $i = 1$, $\bar{i} = 2$).

Algorithm 1 considers each local set $\mathcal{F}_i$. Then, each XFD $f = (C, (X \to B))$ in $\mathcal{F}_i$ is checked and added to $cover\mathcal{F}$ when one of the following properties holds:
(i) There is no path in the source $\bar{i}$ equivalent to the right-hand side of f (line 5). Thus, documents in $\bar{i}$ do not violate f.
(ii) There is no set of paths in the source $\bar{i}$ equivalent to the set on the left-hand side of f (line 7). Since no set of paths in the source $\bar{i}$ correspond to X, no document in $\bar{i}$ violates f.
(iii) In the source $\bar{i}$, there is a path equivalent to C/B that belongs to the closure of a set of paths equivalent to C/X (line 9). Therefore, XFD f exists in both sources and can be added to $cover\mathcal{F}$.

From line 12 to 18, Algorithm 1 takes the fact into account that working with $\mathcal{F}_i$, some XFD in $\mathcal{F}_i^+$ may be neglected. To understand this problem, let us consider sets $\mathcal{F}_1$ and $\mathcal{F}_2$ from which we can derive an XFD $f = (C, (X \to B))$ by different derivation sequences. Suppose that in $\mathcal{F}_1$ we have $f_1, \ldots, f_k, \ldots, f$ while in $\mathcal{F}_2$ we have $f'_1, \ldots, f'_k, \ldots, f$. Moreover, we assume that, due to conditions stated in lines 5, 7 and 9, the dependencies f_k and f'_k are not included in $\mathcal{F}$ and, thus, the derivation of f is not possible from the new set $cover\mathcal{F}$ built by Algorithm 1. This would be a mistake, since f is derived by both $\mathcal{F}_1$ and $\mathcal{F}_2$. One solution would be to start with (in line 4) the closure of $\mathcal{F}_1$ and $\mathcal{F}_2$. However, this solution implies the generation of a too big and, thus, not manipulable set of XFD. Algorithm 1 does better: when the test in line 9 fails, it computes all XFD $f_j = (C, (Y \to A))$ such that:

(*i*) $C/X \in (C, Y)^+$ and $A = B$ or
(*ii*) $C/Y = C/X$ and $C/A \in (C, B)^+$ or
(*iii*) $C/A = C/B$ and f_j is obtained by using Axiom $A4$ on f or
(*iv*) $Y = X \cup Y_1$ and f_j is obtained by using Axiom $A2$ on f to obtain $f' = (C, (X, Y_1 \to B, Y_1))$, and then using Axiom $A3$ on f' and $f'' = (C, (B, Y_1 \to A)) \in G$.

Tests from lines 12-18 are then performed on these computed XFD. In this way, we do not compute the entire closure of a set $\mathcal{F}_i$ but, when necessary, we calculate a part of it. This computation is done by using *closure1Step* and *inverseClosure1Step*. The following example illustrates the computation performed in lines 12-18 of Algorithm 1.

Example 6. Let $\mathcal{F}_1 = \{(C, (A \to B)), (C, (B \to M)), (C, (M \to D)), (C, (D \to E)), (C, (O \to Z))\}$ and let $\mathcal{F}_2 = \{(C, (A \to B)), (C, (B \to M)), (C, (B \to O)), (C, (O \to E)), (C, (D \to N))\}$. Without lines 12-18 in Algorithm 1, the XFD $(C, (A \to E))$, derivable from both $\mathcal{F}_1$ and $\mathcal{F}_2$, would not be derived from $cover\mathcal{F}$.

Let us consider part of the execution of Algorithm 1. Table 4 shows the XFD we obtain when considering each XFD in $\mathcal{F}_1$ (line 3 of Algorithm 1). The first

---

Algorithm 1. Computation of $cover\mathcal{F}$ (set of XFD ensuring the interoperability of S *w.r.t.* S_1 and S_2)

Input:
  - A set of XFD $\mathcal{F}_1$ for schema $\mathcal{D}_1$
  - A set of XFD $\mathcal{F}_2$ for schema $\mathcal{D}_2$
  - The set of paths $\mathbb{P}_1, \mathbb{P}_2$ specified by $\mathcal{D}_1$ and $\mathcal{D}_2$
  - Translation functions Φ_1 and Φ_2

Output: The set of XFD $cover\mathcal{F}$ for the integrated system
1: $cover\mathcal{F} = \emptyset$
2: **for** $i = 1$ **to** 2 **do**
3:     $G = \mathcal{F}_i$
4:     **for each** $(C, (X \rightarrow B)) \in G$ **do**
5:         **if** $\Phi_{\bar{i}}(C/B) \notin \mathbb{P}_{\bar{i}}$ **then**
6:             $cover\mathcal{F} = cover\mathcal{F} \cup \{(C, (X \rightarrow B))\}$
7:         **else if** $\Phi_{\bar{i}}(C/X) \not\subseteq \mathbb{P}_{\bar{i}}$ **then**
8:             $cover\mathcal{F} = cover\mathcal{F} \cup \{(C, (X \rightarrow B))\}$
9:         **else if** $\Phi_{\bar{i}}(C/B) \in \Phi_{\bar{i}}(C, X)^{+}_{\mathcal{F}_{\bar{i}}}$ **then**
10:            $cover\mathcal{F} = cover\mathcal{F} \cup \{(C, (X \rightarrow B))\}$
11:        **else**
12:            $H = closure1Step(C, B, \mathcal{F}_i) \setminus \{C/B\}$
13:            $G = G \cup \{(C, (X \rightarrow D)) \mid C/D \in H\}$
14:            $K = inverseClosure1Step(C, X, \mathcal{F}_i) \setminus \{C/X\}$
15:            $G = G \cup \{(C, (Y \rightarrow B)) \mid C/Y \in K\}$
               % Recall that C/Y is a shorthand for $\{C/A_1, \ldots, C/A_n\}$ and that K is
               a set of paths sets.

16:            $G = G \cup \{(C, (Z \rightarrow B)) \mid (C, (Z \rightarrow B))$ is obtained by using Axiom $A4$
               on $(C, (X \rightarrow B))\}$
17:            % Notice that Z is a set of prefixes of paths in X or B

18:            $G = G \cup \{(C, (X, W \rightarrow V)) \mid (C, (X, W \rightarrow V))$ is obtained by using
               the Axioms $A2, A3$ on $(C, (X \rightarrow B))$ and $(C, (B, W \rightarrow V))$ where
               $(C, (B, W \rightarrow V)) \in G\}$
19:        **end if**
20:     **end for**
21: **end for**
22: **return** $cover\mathcal{F}$

---

column of this table shows the XFD in G being verified. The second column indicates XFD that are added to G due to lines 12-18. Finally the last column shows XFD that are inserted in $cover\mathcal{F}$.

Table 4 is obtained by following the execution of Algorithm 1. For instance, let us consider the third line in Table 4: the case when the XFD $(C, (M \rightarrow D))$ in $\mathcal{F}_1$ is taken in line 4 of Algorithm 1. This XFD does not verify any condition among conditions in lines 5, 7 and 9. When line 12 is executed, the set $H = \{C/E\}$ is computed, since $closure1Step(C, D, \mathcal{F}_1)$ gives $\{C/D, C/E\}$. Thus, the XFD $(C, (M \rightarrow E))$ is added to G (line 13). When line 14 is executed,

Table 4. Computation of (part) of $cover\mathcal{F}$: XFD obtained when considering $\mathcal{F}_1$

G (**XFD being considered**)	**Add to G**	**XFD added to $cover\mathcal{F}$**
$(C, (A \rightarrow B))$		$(C, (A \rightarrow B))$ (cond. line 9)
$(C, (B \rightarrow M))$		$(C, (B \rightarrow M))$ (cond. line 9)
$(C, (M \rightarrow D))$	$(C, (M \rightarrow E))$ $(C, (B \rightarrow D))$ $(C, ([] \rightarrow D))$	
$(C, (D \rightarrow E))$	$(C, (M \rightarrow E))$ $(C, ([] \rightarrow E))$	
$(C, (O \rightarrow Z))$		$(C, (O \rightarrow Z))$ (cond. line 5)
$(C, (M \rightarrow E))$	$(C, (B \rightarrow E))$ $(C, ([] \rightarrow E))$	
$(C, (B \rightarrow D))$	$(C, (B \rightarrow E))$ $(C, (A \rightarrow D))$ $(C, ([] \rightarrow D))$	
$(C, (B \rightarrow E))$		$(C, (B \rightarrow E))$ (cond. line 9)
$(C, (A \rightarrow D))$	$(C, (A \rightarrow E))$ $(C, ([] \rightarrow D))$	
$(C, (A \rightarrow E))$		$(C, (A \rightarrow E))$ (cond. line 9)

the set $K = \{\{C/B\}\}$ is computed, since $inverseClosure1Step(C, M, \mathcal{F}_1)$ gives $\{\{C/B\}, \{C/M\}\}$. Thus, the XFD $(C, (B \rightarrow D))$ is added to G (line 15). When line 16 is executed, the XFD $(C, ([] \rightarrow D))$ is added to G. Notice that these three XFD are analysed later (lines 6 and 7 of Table 4). They are not included in $cover\mathcal{F}$, but generate other XFD as, for instance, $(C, (A \rightarrow E))$, which is finally added to $cover\mathcal{F}$. □

7.2   Properties of $cover\mathcal{F}$

In this section we prove that Algorithm 1 works correctly, and fulfills our goals. First we introduce Lemma 2, telling us which XFD should be added to the set $\mathcal{F} \setminus \{f\}$ in order to ensure the derivation of $\mathcal{F}^+$, except for f. Indeed, the derivation of f from the new set G is neither guaranteed nor proscribed.

Lemma 2. Let $\mathcal{F}$ be a set of XFD such that $(C, (X \rightarrow Y)) \in \mathcal{F}$. Let $(C, (Z_1 \rightarrow Z_2))$ be an XFD different from $(C, (X \rightarrow Y))$. If $\mathcal{F} \vdash (C, (Z_1 \rightarrow Z_2))$ then $G \vdash (C, (Z_1 \rightarrow Z_2))$ where G is obtained from $\mathcal{F}$ as follows:

$$G = \mathcal{F} \cup \mathcal{F}_1 \cup \mathcal{F}_2 \cup \mathcal{F}_3 \cup \mathcal{F}_4 \setminus \{(C, (X \rightarrow Y))\}$$

where $\mathcal{F}_1 = \{(C, (X \rightarrow V)) \mid V \in closure1Step(C, Y, \mathcal{F})\}$,
$\mathcal{F}_2 = \{(C, (W \rightarrow Y)) \mid W \in inverseClosure1Step(C, X, \mathcal{F})\}$,
$\mathcal{F}_3 = \{(C, (\{P'_1, \ldots, P'_n\} \rightarrow Y)) \mid \{P'_1, \ldots, P'_n\}$ respects conditions for applying Axiom A4 on $(C, (X \rightarrow Y))\}$,
$\mathcal{F}_4 = \{(C, (X, W \rightarrow V)) \mid (C, (Y, W \rightarrow V)) \in \mathcal{F}\}$. □

Sketch of proof: Since $\mathcal{F} \vdash (C, (Z_1 \rightarrow Z_2))$, there exists a sequence α of XFD containing XFD in $\mathcal{F}$ such that α derives $(C, (Z_1 \rightarrow Z_2))$. The crucial point of the proof is when we suppose that α contains $(C, (X \rightarrow Y))$. Thus, α has sub-sequences for which one of the following conditions holds:

- it derives the set of paths X in one step or
- it derives the paths $Y_1, \ldots, Y_n \in closure1Step(C, Y, \mathcal{F})$ or
- it derives path Y in one step by using Axiom A4 on $(C, (X \rightarrow Y))$ or
- it derives path V from X, W where $(C, (Y, W \rightarrow V)) \in \mathcal{F}$.

The proof consists in replacing the XFD $(C, (X \rightarrow Y))$ and all sub-sequences of α respecting the above conditions, by some of the new XFD which are added to $\mathcal{F}$ for obtaining G. By considering G and the new derivation sequence (obtained after replacing XFD in α) we can derive $(C, (Z_1 \rightarrow Z_2))$.   □

Now, given two sets of XFD, $\mathcal{F}_i$ and $\mathcal{F}_{\bar{i}}$, we define set $\mathcal{K}_i$ of XFD which contains all the XFD f which can be obtained from $\mathcal{F}_i$ but that cannot be violated by documents in $X_{\bar{i}}$ due to one of the two reasons:
(a) the right-hand side of f is a path B which belongs to $\mathbb{P}_i$ but not to $\mathbb{P}_{\bar{i}}$ or
(b) the left-hand side of f is a set of paths X which is included in $\mathbb{P}_i$ but not in $\mathbb{P}_{\bar{i}}$.

Formally, we have $\mathcal{K}_i = \{X \rightarrow A \mid X \rightarrow A \in \mathcal{F}_i^+$ and $[((X \subseteq \mathbb{P}_i)$ and $(X \not\subseteq \mathbb{P}_{\bar{i}}))$ or $(A \in (\mathbb{P}_i \setminus \mathbb{P}_{\bar{i}}))]\}$.

In [2] we show an algorithm, starting with $\mathcal{F}_1^+$ and $\mathcal{F}_2^+$, instead of $\mathcal{F}_1$ and $\mathcal{F}_2$, that computes the set $\mathcal{F} = (\mathcal{F}_1^+ \cap \mathcal{F}_2^+) \cup \mathcal{K}_1 \cup \mathcal{K}_2$. We prove some properties of $\mathcal{F}$. This set $\mathcal{F}$ is the biggest set of XFD that does not violate any document in X_i and $X_{\bar{i}}$.

Theorem 1. *The set cover$\mathcal{F}$, returned by Algorithm 1, is equivalent to (or is a cover of) the set of XFD $\mathcal{F} = (\mathcal{F}_1^+ \cap \mathcal{F}_2^+) \cup \mathcal{K}_1 \cup \mathcal{K}_2$ (cover$\mathcal{F} \equiv \mathcal{F}$).*   □

Sketch of proof: For proving that $cover\mathcal{F} \equiv \mathcal{F}$, we will prove that: $(A1)$ $\forall f \in cover\mathcal{F}$, $\mathcal{F} \vdash f$ and $(A2)$ $\forall f \in \mathcal{F}$, $cover\mathcal{F} \vdash f$.

$(A1)$ By the following Algorithm 1, we can easily prove that each XFD added to $cover\mathcal{F}$ is also in $\mathcal{F}$. Thus, we have $cover\mathcal{F} \subseteq \mathcal{F}$ which is stronger than just proving that $\mathcal{F} \vdash f$ for any XFD $f \in cover\mathcal{F}$.

$(A2)$ Let $f = (C, (Y \rightarrow A))$ be an XFD in $\mathcal{F}_1^+ \cap \mathcal{F}_2^+$. Thus, we know that $C/A \in (C, Y)_{\mathcal{F}_1}^+$ and $C/A \in (C, Y)_{\mathcal{F}_2}^+$. Since $C/A \in (C, Y)_{\mathcal{F}_1}^+$, there is a derivation sequence $\alpha = f_1, \ldots, f_n$ which derives f. If $cover\mathcal{F}$ contains all the XFD of $\mathcal{F}_1$ taking part in α then we have $cover\mathcal{F} \vdash f$. Otherwise there is at least one XFD of $\mathcal{F}_1$ (denote it by f_k) that takes part in α but does not belong to $cover\mathcal{F}$. Since $f_k \notin cover\mathcal{F}$ then from lines 12-16, we know that f_k is deleted from G and that some other XFD h is inserted in G. By using Lemma 2, we have $G \vdash f$. If all new functional dependencies h satisfy conditions in lines 5, 7, 9 they are added to $cover\mathcal{F}$. Otherwise, they are analysed in lines 12-16 and the process goes on until f is added to G and, thus, to $cover\mathcal{F}$. With the same arguments we can prove that $f \in cover\mathcal{F}^+$ when $f \in \mathcal{K}_1$ or $f \in \mathcal{K}_2$.   □

7.3   Complexity of Our Method

Algorithm 1 depends on the algorithm that computes the closure of a set of paths $((C, X[E])^+)$, and on the algorithm that computes just one step of the inverse closure of a set of paths.

The running time of the closure algorithm, in the worst (unlikely) case, is $O(|\mathbb{P}|^2 \cdot (|f| \cdot |\mathcal{F}| + |\mathbb{P}|))$ where $|\mathcal{F}|$ is the cardinality of $\mathcal{F}$ and $|f|$ is the size of the longest XFD in $\mathcal{F}$. The running time of the $inverseClosure1Step$ algorithm is $O((|f|^n \cdot |\mathcal{F}|)^{|X|})$ where $|f|$ is the size of the longest XFD in $\mathcal{F}$, n is the number of paths on the left-hand side of f and $|X|$ is the cardinality of the set of paths X on which the function $inverseClosure1Step$ is performed.

In the worst case, Algorithm 1 will treat about $|\mathcal{F}_i| \cdot |\mathbb{P}_i|$ functional dependencies for each set $\mathcal{F}_i$. The worst case occurs when for each XFD $f = (C, (X \to P))$ in $\mathcal{F}_i$, $(C, X)^+$ contains $|\mathbb{P}_i|$ paths and just one path is added to $(C, X)^+$ in each step of the loop of the closure algorithm and no XFD is added to $cover\mathcal{F}$. Hence, in this case, lines 12-18 of Algorithm 1 will be executed $|\mathbb{P}_i|$ times for each XFD in $\mathcal{F}_i$. The complexity of Algorithm 1 is $O(|\mathcal{F}_i| \cdot |\mathbb{P}_i| \cdot (g + h))$ where g is the complexity of the closure algorithm and h is the complexity of the $inverseClosure1Step$ algorithm. The variables that are determinants in the complexity are the cardinality of $\mathcal{F}$ and $\mathbb{P}$. In practice $|X|$ and n are not greater than 5 and, thus, have little importance when compared with the size of $\mathcal{F}$ and $\mathbb{P}$.

8   Experimental Results

In order to examine the performance of Algorithm 1, we run several experiments on synthetic data. Algorithm 1 has as input two local systems $S_1 = (\mathcal{D}_1, \mathcal{F}_1, O)$ and $S_2 = (\mathcal{D}_2, \mathcal{F}_2, O)$, and computes the set $cover\mathcal{F}$ which contains only the XFD for which no violation is possible when considering document sets from S_1 and S_2. Recall that we assume the existence of a pre-processing step where the correspondence among paths on the different local documents is established. This pre-processing step is built on the basis of an ontology alignment but it is out of purpose in this paper. In this section we assume that this correspondence has already been done: paths are represented on the basis of a common ontology O.

We take into account two parameters in the experiments: (i) the number of paths obtained from $\mathcal{D}_1$ and $\mathcal{D}_2$, and (ii) the number of XFD in $|\mathcal{F}_1| + |\mathcal{F}_2|$.

Tree $\mathcal{T}$ (Figure 3) guides the way we perform our experiments. $\mathcal{T}$ is built by repeating the pattern tree in Figure 3 several times. To perceive the difference between the sub-trees of $\mathcal{T}$, we relabel the nodes of the pattern tree by adding the index k ($k \geq 1$). We say *sub-tree k* to refer to the kth tree pattern in $\mathcal{T}$. For example, in Figure 3, $A_{1,1}$ refers to element A_1 of subtree $k = 1$ while and $A_{1,2}$ refers to element A_1 of subtree $k = 2$.

Our experiments consist in generating $cover\mathcal{F}$ from sets $\mathcal{F}_1$ and $\mathcal{F}_2$ which increase at each test by assuming the existence of bigger sets of paths $\mathbb{P}_1$ and $\mathbb{P}_2$ and, therefore, larger trees $\mathcal{T}$. In the text, we usually refer to tree $\mathcal{T}$ to indicate the type of documents (the schema) we are dealing with. In this context, let us define $\mathbb{P}_1^j$ as the set of paths containing all the paths in the tree $\mathcal{T}$ except the paths $C/R_{1,k}/G_{1,k}$ (with $k \leq j$), and $\mathbb{P}_2^j$ as the set of paths containing all the paths in the tree $\mathcal{T}$ except the paths $C/R_{1,k}/F_{1,k}$ (with $k \leq j$). We suppose that the set of paths $\mathbb{P}_1^j$ (respectively $\mathbb{P}_2^j$) is generated from $\mathcal{D}_1$ (respectively $\mathcal{D}_2$).

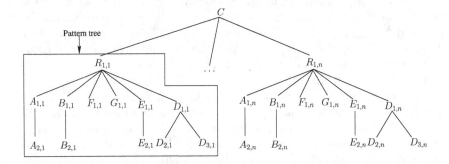

Fig. 3. Tree $\mathcal{T}$ built by repeating n times the pattern tree

The set of XFD $\mathcal{F}_1^j$ (respectively $\mathcal{F}_2^j$) is defined over paths in $\mathbb{P}_1^j$ (respectively $\mathbb{P}_2^j$). Table 5 shows the XFD in $\mathcal{F}_1^j$ and $\mathcal{F}_2^j$. Sets $\mathcal{F}_1^j$ and $\mathcal{F}_2^j$ contain both XFD (1) and (2). However, XFD (3a), (4a) and (5a) are only in $\mathcal{F}_1^j$ and XFD (3b), (4b) and (5b) are only in $\mathcal{F}_2^j$. With XFD (4a) and (5a), we can derive the XFD (6) $(C/R_{1,k}, (\{A_{1,k}/A_{2,k}, B_{1,k}/B_{2,k}\} \rightarrow E_{1,k}/E_{2,k}))$ and with XFD (4b) and (5b), we can also derive the XFD (6). Hence, $\mathcal{F}_1^j$ and $\mathcal{F}_2^j$ derive XFD (6) but by different ways. We can remark that $|\mathcal{F}_1^1| = |\mathcal{F}_2^1| = 5$, $|\mathcal{F}_1^2| = |\mathcal{F}_2^2| = 10$ and $\mathcal{F}_1^1 \subset \mathcal{F}_1^2$, $\mathcal{F}_2^1 \subset \mathcal{F}_2^2$.

Table 5. Contents of the XFD sets $\mathcal{F}_1^j$ and $\mathcal{F}_2^j$ used in the experiments

$\mathcal{F}_1^j$	$\mathcal{F}_2^j$
(1) $(C/R_{1,k}, (\{A_{1,k}, B_{1,k}\} \rightarrow D_{1,k}))$	(1) $(C/R_{1,k}, (\{A_{1,k}, B_{1,k}\} \rightarrow D_{1,k}))$
(2) $(C/R_{1,k}, (\{D_{1,k}\} \rightarrow E_{1,k}))$	(2) $(C/R_{1,k}, (\{D_{1,k}\} \rightarrow E_{1,k}))$
(3a) $(C/R_{1,k}, (\{E_{1,k}\} \rightarrow F_{1,k}))$	(3b) $(C/R_{1,k}, (\{E_{1,k}\} \rightarrow G_{1,k}))$
(4a) $(C/R_{1,k}, (\{A_{1,k}/A_{2,k}, B_{1,k}/B_{2,k}\} \rightarrow$ $D_{1,k}/D_{2,k}))$	(4b) $(C/R_{1,k}, (\{A_{1,k}/A_{2,k}, B_{1,k}/B_{2,k}\} \rightarrow$ $D_{1,k}/D_{3,k}))$
(5a) $(C/R_{1,k}, (\{D_{1,k}/D_{2,k}\} \rightarrow E_{1,k}/E_{2,k}))$	(5b) $(C/R_{1,k}, (\{D_{1,k}/D_{3,k}\} \rightarrow E_{1,k}/E_{2,k}))$

The algorithm was implemented in Java and the tests have been done on an Intel Quad Core i3-2310M with 2.10GHz and 8GB of memory. We have used three scenarios for performing our tests.

In the first scenario we examine the influence of the size of $\mathcal{F}_1$ and $\mathcal{F}_2$ on the execution time of Algorithm 1. We have used $\mathcal{F}_1^j$ and $\mathcal{F}_2^j$, such that $1 \leq j \leq 45$. Figure 4 shows reasonable execution time (approximately 2 minutes) for computing $cover\mathcal{F}$ from sets of XFD $\mathcal{F}_1$ and $\mathcal{F}_2$ where $|\mathcal{F}_1| + |\mathcal{F}_2| = 450$. Figure 4 also shows how $cover\mathcal{F}$ increases: at each step as we add 25 XFD to $|\mathcal{F}_1| + |\mathcal{F}_2|$, set $cover\mathcal{F}$ has about 50 XFD more than its previous version.

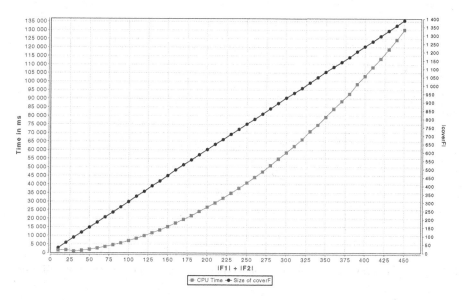

Fig. 4. Scenario 1: CPU time for the computing of $cover\mathcal{F}$ and the evolution of its size

In the second scenario we examine again the influence of the size of $\mathcal{F}_1$ and $\mathcal{F}_2$ on the execution time of Algorithm 1. Notice that, in the first scenario, the functional dependencies involving index k concerns only one subtree. In this second scenario, we allow an XFD involving index $k = 1$ to derive an XFD involving index $k = 2$, and so on. To do this, we add to $\mathcal{F}_1^j$ (resp. $\mathcal{F}_2^j$) the XFD of the form $(7a)$ $(C, (\{R_{1,k}/E_{1,k-1}/E_{2,k-1}\} \rightarrow R_{1,k}/D_{1,k}/D_{2,k}))$, resp. $(7b)$ $(C, (\{R_{1,k}/E_{1,k-1}/E_{2,k-1}\} \rightarrow R_{1,k}/D_{1,k}/D_{3,k}))$, with $2 \leq k \leq j$.

As shown in Figure 5, the execution time for computing $cover\mathcal{F}$ is more important than the one obtained with the first scenario. For instance, for sets $\mathcal{F}_1$ and $\mathcal{F}_2$ (such that $|\mathcal{F}_1| + |\mathcal{F}_2| = 262$) we need 53 minutes to compute $cover\mathcal{F}$. This behaviour is explained by two facts:

- XFD of the form (7a) and (7b) are not added to $cover\mathcal{F}$ due to the condition in line 9 of Algorithm 1. Checking this condition is an expensive task because the computation of $(C, R_{1,k}/E_{1,k-1}/E_{2,k-1})^+_{\mathcal{F}_i}$ involves many paths.
- For this example, lines 12-15 of Algorithm 1 generate many XFD dramatically increasing the number of XFD in $cover\mathcal{F}$. Indeed, $|cover\mathcal{F}|$ has about 10610 XFD when $|\mathcal{F}^j_1| + |\mathcal{F}^j_2|$ is 262.

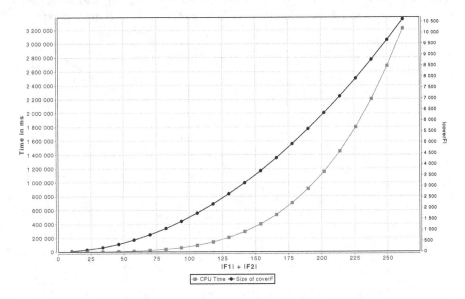

Fig. 5. Scenario 2: CPU time for the computing of $cover\mathcal{F}$ and the evolution of its size

In the third scenario, we compare the algorithm built to compute $\mathcal{F} = (\mathcal{F}^+_1 \cap \mathcal{F}^+_2) \cup \mathcal{K}_1 \cup \mathcal{K}_2$ from $\mathcal{F}^+_1$ and $\mathcal{F}^+_2$ (presented in [2]) with Algorithm 1 (which computes $cover\mathcal{F}$). Recall that in Section 7, we have shown that $cover\mathcal{F}$ is equivalent to $\mathcal{F}$. Now, Table 6 compares these two algorithms. Line 1 in Table 6 shows the results with sets $\mathcal{F}^1_1$ and $\mathcal{F}^1_2$ while line 4 shows the result with $\mathcal{F}^2_1$ and $\mathcal{F}^2_2$, and line 5 shows the result with $\mathcal{F}^3_1$ and $\mathcal{F}^3_2$, the same sets used in scenario 1. When computing the set $\mathcal{F}$ for sets $\mathcal{F}^3_1$ and $\mathcal{F}^3_2$ with the algorithm in [2], we obtain an *out-of-memory* error after 5 minutes. For the same sets of XFD, Algorithm 1 takes approximately 2.9 seconds and $|cover\mathcal{F}| = 92$. Since the test concerning line 5 does not produce a result for the algorithm in [2], we perform tests of line 2 and 3 on a modified tree, *i.e.*, on $\mathcal{T}$ without the leaves. In other words, we delete nodes $A_{2,2}, B_{2,2}, D_{2,2}, D_{3,2}$ and $E_{2,2}$ from a tree $\mathcal{T}$ with $k = 2$ sub-trees. The tree considered in line 3 contains nodes $F_{1,2}$ and $G_{1,2}$ in addition to nodes in the tree considered in line 2. As expected, in all cases, Algorithm 1 is much more efficient than the algorithm in [2]. Moreover, the size of $\mathcal{F}$ grows dramatically while the size of $cover\mathcal{F}$ increases slightly.

Table 6. Comparison: t_1 is the time needed to compute $\mathcal{F}$ and t_2 is the time needed to compute $cover\mathcal{F}$

	$|\mathbb{P}_1| + |\mathbb{P}_2|$	$|\mathcal{F}_1|$	$|\mathcal{F}_2|$	$|\mathcal{F}_1^+|$	$|\mathcal{F}_2^+|$	$|\mathcal{F}|$	t_1 (ms)	$|cover\mathcal{F}|$	t_2 (ms)
1	12	5	5	3 835	3 835	5 755	19 211	30	160
2	17	7	7	3 865	3 865	5 785	24 401	32	195
3	18	8	8	3 928	3 928	5 911	26 476	34	204
4	23	10	10	7 670	7 670	11 510	165 460	61	503
5	34	15	15	?	?	?	> 5min	92	2 949

Our experiments confirm the time complexity presented in Section 7.3, and reinforce the importance of computing the smaller set $cover\mathcal{F}$ instead of the equivalent set $\mathcal{F}$ considered in [2]. The worst case happens when documents have many equivalent paths and derivations that involve a lot of paths. We have used the closure algorithm to test, successfully, the equivalence between $\mathcal{F}$ and $cover\mathcal{F}$ on several examples. These tests contribute to the validation of the correctness of our method.

9   Conclusions

We are motivated by applications on a multiple system environment and we have presented a method for establishing the biggest set of XFD that can be satisfied by any document conceived to respect local XFD. One important originality of our work is the fact that we do not deal with data, only with the available constraints (XFD in our case). Our approach is not only interesting for multiple system applications, but also in a conservative constraint evolution perspective. To reach our goals, a new axiom system, built for XFD defined over a context and two kinds of equality, was introduced and proved to be sound and complete.

As some future directions that follow from this work, we mention:

- By using the schema evolution method of [9] together with our computation of $cover\mathcal{F}$, the generation of a new type and a new set of integrity constraints that will allow interoperability without abolishing constraint verification. We are currently working on a platform that puts together these tools.
- The extension of our method to other kinds of integrity constraints such as inclusion constraints.
- An incremental computation of $cover\mathcal{F}$, following the evolution of local constraints or systems.
- The detection of local XFD that are not selected in $cover\mathcal{F}$ but that could be included in it by correcting the associated documents that do not respect them.
- The implementation of an XFD validator over the local systems, in a map-reduce approach, by considering the set $cover\mathcal{F}$ as the set of constraints that should be respected by the data of our multiple system.

A   Soundness of the Axiom System

In this section we prove that our axiom system is sound, *i.e.*, our axioms always lead to true conclusions when we deal with complete XML trees. We start by proving some lemmas. The first one deals with properties concerning the longest common prefix of paths. The following example illustrates the situation it concerns.

Example 7. We consider the XML document in Figure 2 and the following paths: $P_K = univ/undergraduate/@domain$, $P_J = univ/undergraduate/students/-student/idSt$ and $P_I = univ/undergraduate/students/student/nameSt$. In this situation we have $P_I \cap P_J = univ/undergraduate/students/student$ and $P_J \cap P_K = univ/undergraduate$. Clearly, $P_J \cap P_K \preceq P_I \cap P_J$. Then, consider path instances where $isInst\_lcp(P_I, I, P_J, J) = true$ and $isInst\_lcp(P_I, I, P_K, K) = true$. For instance, let instance $K = \epsilon/2/2.1$, instance $J = \epsilon/2/2.2/-2.2.0/2.2.0.0$ and $I = \epsilon/2/2.2/2.2.0/2.2.0.1$. Notice that in this case we also have: $isInst\_lcp(P_J, J, P_K, K) = true$.   □

The above example suggests that a kind of transitivity property could be established for the function $isInst\_lcp$. The following lemma proves that this is actually possible.

Lemma 3. Let $\mathcal{T}$ be an XML document and $\mathbb{P}$ its associated set of simple paths. Let P_I, P_J, P_K be distinct paths in $\mathbb{P}$. If $P_J \cap P_K \preceq P_I \cap P_J$ and $isInst\_lcp(P_I, I, P_J, J) = isInst\_lcp(P_I, I, P_K, K) = true$ then $isInst\_lcp(P_J, J, P_K, K) = true$.   □

The next example illustrates a special situation where an XFD not satisfied by a given document has at the left-hand side a path which is a prefix of the path on the right-hand side.

Example 8. We consider the example in Figure 2, the XFD $f = (univ/under-gradute, (\{courses//titleC, courses//prerequisitesC\} \rightarrow courses//pre-requisitesC/codeC))$ and the branching path M defined by f. The document of Figure 2 does not satify f. Notice that $P_2 = univ/undergradute/courses/course/prerequisitesC$ in the left-hand side of f is a prefix for $P = univ/undergradute/courses/course/prerequisitesC/codeC$ in right-hand side of f. Also remark that we can find two projections of the XML tree over M such that $Last(\Pi_M^1(\mathcal{T})[C/P_2]) =_N Last(\Pi_M^2(\mathcal{T})[C/P_2])$: the two projections ending on node 2.3.0.2.   □

The following lemma proves that in situations as the one illustrated by Example 8 we can always find two projections of the XML tree over the branching path M such that $Last(\Pi_M^1(\mathcal{T})[C/P_j]) =_N Last(\Pi_M^2(\mathcal{T})[C/P_j])$, where P_j is the path on the left-hand side which is a prefix of the one on the right-hand side.

Lemma 4. Let $\mathcal{T}$ be an XML document, $f = (C, (\{P_1 [E_1], \ldots, P_n [E_n]\} \to P_{n+1} [E_{n+1}]))$ an XFD and let M be the branching path $\{C/P_1, \ldots, C/P_{n+1}\}$. If $\mathcal{T} \not\models f$ and there exists a $j \in [1 \ldots n]$ such that $P_j \preceq P_{n+1}$ then we can find two projections $\Pi_M^1(\mathcal{T})$ and $\Pi_M^2(\mathcal{T})$ for M in $\mathcal{T}$ such that $Last(\Pi_M^1(\mathcal{T})[C/P_j]) =_N Last(\Pi_M^2(\mathcal{T})[C/P_j])$. $\square$

In this appendix we just show the soundness proof of axiom A4.

Theorem 2. *Axiom A4 is sound for XFD on complete XML trees.* $\square$

Proof: We consider a complete XML document $\mathcal{T} = (t, type, value)$.
A4: Let $f = (C, (\{P_1' [E_1'], \ldots, P_n' [E_n']\} \to P_{n+1} [E_{n+1}]))$ and $f' = (C, (\{P_1 [E_1], \ldots, P_n [E_n]\} \to P_{n+1} [E_{n+1}]))$. The proof is by contradiction. Suppose that $\mathcal{T} \models f$ but $\mathcal{T} \not\models f'$. From Axiom A1, we can assume that for all $i \in [1 \ldots n]$, $P_i \neq P_{n+1}$. From Definition 5, we can deduce that there exist two projections $\Pi_M^1(\mathcal{T})$ and $\Pi_M^2(\mathcal{T})$ for the branching path $M = \{C/P_1, \ldots, C/P_{n+1}\}$ in $\mathcal{T}$ such that $\tau^1[C/P_{n+1}] \neq_{E_{n+1}} \tau^2[C/P_{n+1}]$ and $\tau^1[C/P_1, \ldots, C/P_n] =_{E_i, i \in [1 \ldots n]} \tau^2[C/P_1, \ldots, C/P_n]$. We now show that there exist two projections $\Pi_{M'}^1(\mathcal{T})$ and $\Pi_{M'}^2(\mathcal{T})$, constructed from $\Pi_M^1(\mathcal{T})$ and $\Pi_M^2(\mathcal{T})$, for the branching path $M' = \{C/P_1', \ldots, C/P_n', C/P_{n+1}\}$ in $\mathcal{T}$ such that:

$$u^1[C/P_1', \ldots, C/P_n'] =_{E_i', i \in [1 \ldots n]} u^2[C/P_1', \ldots, C/P_n'] \quad \text{and} \tag{1}$$

$$u^1[C/P_{n+1}] \neq_{E_{n+1}} u^2[C/P_{n+1}]. \tag{2}$$

However, from our hypothesis we know that for all two projections $\Pi_{M'}^1(\mathcal{T})$ and $\Pi_{M'}^2(\mathcal{T})$ such that (1) is satisfied then we have $u^1[C/P_{n+1}] =_{E_{n+1}} u^2[C/P_{n+1}]$. If $\Pi_{M'}^1(\mathcal{T})$ and $\Pi_{M'}^2(\mathcal{T})$ really exist, we have a contradiction with (2) and the axiom A4 will be satisfied.

The proof is by showing that it is possible to obtain two projections for M' satisfying (1) and (2). We start by considering that $\Pi_{M'}^1(\mathcal{T})[C/P_i] = \Pi_M^1(\mathcal{T})[C/P_i]$ and $\Pi_{M'}^2(\mathcal{T})[C/P_i]) = \Pi_M^2(\mathcal{T})[C/P_i] \; \forall i \in [1 \ldots n+1]$.

1. If $\exists k \in [1 \ldots n]$ such that $P_k \preceq P_{n+1}$ (Figure 6(a)) then, from Lemma 4, we can consider that :

$$Last(\Pi_{M'}^1(\mathcal{T})[C/P_k]) =_N Last(\Pi_{M'}^2(\mathcal{T})[C/P_k]). \tag{3}$$

Since t is complete there exist instances J_i such that $\forall i \in [1 \ldots n]$, $\Pi_{M'}^1(\mathcal{T})[C/P_i] \preceq J_i$ and $J_i \in Instances(C/P_i', t)$ (see Figure 6(a)). Let $\forall i \in [1 \ldots n]$, $\Pi_{M'}^1(\mathcal{T})[C/P_i'] = \Pi_{M'}^2(\mathcal{T})[C/P_i'] = J_i$. Then by considering these instances J_i for paths C/P_i' and by using Lemma 3, we can show that $\forall i, j \in [1 \ldots n+1]$ (recall that we consider that $P_{n+1}' = P_{n+1}$):

$$isInst\_lcp(C/P_i', \Pi_{M'}^1(\mathcal{T})[C/P_i'], C/P_j', \Pi_{M'}^1(\mathcal{T})[C/P_j']) = true \tag{4}$$

$$\text{and } isInst\_lcp(C/P_i', \Pi_{M'}^2(\mathcal{T})[C/P_i'], C/P_j', \Pi_{M'}^2(\mathcal{T})[C/P_j']) = true. \tag{5}$$

Thus, in this case, it is possible to have projections $\Pi_{M'}^1(\mathcal{T})$ and $\Pi_{M'}^2(\mathcal{T})$ satisfying 1 and 2.

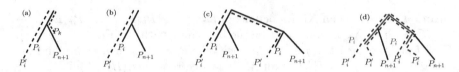

Fig. 6. Graphical representation of paths and possible projections. Case (a) Path $P_k \prec P_{n+1}$ where P_k is one of the paths in the left-handside of XFD f' and $P_i \preceq P'_i$. Case (b): Node equality for last nodes in P_i. Both projections $\Pi^1_{M'}(\mathcal{T})$ and $\Pi^2_{M'}(\mathcal{T})$ have the same instance for path P'_i. Case (c) and (d): Value equality for last nodes in P_i. In case (c) there is only one instance of P'_i ($P_i \preceq P'_i$). In case (d) there are two instances of P'_i ($P_i \preceq P'_i$). $\qquad\square$

2. Otherwise if $\forall i \in [1 \ldots n]$, $P_i \not\preceq P_{n+1}$ and $P_i \preceq P'_i$ we can have the following situations:

  (a) If we consider **node equality**, we have $Last(\Pi^1_{M'}(\mathcal{T})[C/P_i]) =_N Last(\Pi^2_{M'}(\mathcal{T})[C/P_i])$ (Figure 6(b)). Since t is complete there exists an instance J_i such that $\Pi^1_{M'}(\mathcal{T})[C/P_i] \preceq J_i$ and $J_i \in Instances(C/P'_i, t)$. Let $\Pi^1_{M'}(\mathcal{T})[C/P'_i] = \Pi^2_{M'}(\mathcal{T})[C/P'_i] = J_i$.

  (b) If we consider **value equality**, we have $Last(\Pi^1_{M'}(\mathcal{T})[C/P_i]) =_V Last(\Pi^2_{M'}(\mathcal{T})[C/P_i])$. Since t is complete there exist instances J^1_i, J^2_i such that $\Pi^1_{M'}(\mathcal{T})[C/P_i] \preceq J^1_i$, $\Pi^2_{M'}(\mathcal{T})[C/P_i] \preceq J^2_i$ and $J^1_i, J^2_i \in Instances(C/P'_i, t)$.
    - If $Last(J^1_i) =_V Last(J^2_i)$ then let $\Pi^1_{M'}(\mathcal{T})[C/P'_i] = J^1_i$ and $\Pi^2_{M'}(\mathcal{T})[C/P'_i] = J^2_i$ (Figure 6(c)).
    - Otherwise if $Last(J^1_i) \neq_V Last(J^2_i)$ then, since $P_i \preceq P'_i$ and $Last(\Pi^1_{M'}(\mathcal{T})[C/P_i]) =_V Last(\Pi^2_{M'}(\mathcal{T})[C/P_i])$, there exists two instances J^3_i, J^4_i such that $\Pi^1_{M'}(\mathcal{T})[C/P_i] \preceq J^3_i$, $\Pi^2_{M'}(\mathcal{T})[C/P_i] \preceq J^4_i$ and $J^3_i, J^4_i \in Instances(C/P'_i, t)$, $Last(J^1_i) =_V Last(J^4_i)$ and $Last(J^2_i) =_V Last(J^3_i)$. In this case, let $\Pi^1_{M'}(\mathcal{T})[C/P'_i] = J^1_i$ and $\Pi^2_{M'}(\mathcal{T})[C/P'_i] = J^4_i$ (Figure 6(d)).

Then by considering these instances J_i for paths C/P'_i and by using Lemma 3, we can show (4) and (5) in each case. Since $\Pi^1_{M'}(\mathcal{T})$ and $\Pi^2_{M'}(\mathcal{T})$ exist and conditions (1), (2) are satisfied, we can conclude that $A4$ is sound.

B   Completeness of the Axiom System

Before tackling the completeness issue, it is important to show the central fact about the closure of a set of paths. It enables us to tell on a glance whether an XFD follows from a set $\mathcal{F}$ by the axiom system. The next lemma tells us how.

Lemma 5. Let $X = \{P_1, \ldots, P_n\}$ and $Y = \{P_{n+1}, \ldots, P_{n+m}\}$ be two sets of paths. Let $E = (E_1, \ldots, E_n)$ and $E' = (E_{n+1}, \ldots, E_{n+m})$. We have $\mathcal{F} \vdash (C, (\{P_1 [E_1], \ldots, P_n [E_n]\} \rightarrow \{P_{n+1} [E_{n+1}], \ldots, P_{n+m} [E_{n+m}]\}))$ iff $C/Y[E'] \subseteq (C, X[E])^+$. $\qquad\square$

To prove the completeness of our axiom system, we would like to define a special tree having two instances (except for the root node) for every path $P \in \mathbb{P}$. However, the following examples show that depending on the conditions imposed on paths, it is not possible to have *two* instances for *every* path $P \in \mathbb{P}$.

Example 9. We want to build a complete tree having exactly two instances for each path in $\mathbb{P}$. Let us consider value equality and two paths P and Q such that $P \prec Q$. We denote by I_{P_1} and I_{P_2} the two instances of P on a tree t. We denote by I_{Q_1} and I_{Q_2} the two instances of Q on a tree t. Suppose that $Last(I_{P_1}) =_V Last(I_{P_2})$ and $Last(I_{Q_1}) \neq_V Last(I_{Q_2})$. Based on this situation, the functional dependency $P \rightarrow Q$ is not satisfied by this tree. Then, we can apply Lemma 4, to conclude that there is an instance of P which is a prefix of both (distinct) instances of Q. As we want just two instances for each path, to have two instances of Q we should have $Last(I_{P_1}) =_N Last(I_{P_2})$. In other words, in this situation, we cannot have a tree with two instances for P. Indeed, Figure 7 illustrates that a tree having two instances of P and respecting the constraints $Last(I_{P_1}) =_V Last(I_{P_2})$ and $Last(I_{Q_1}) \neq_V Last(I_{Q_2})$ must have four instances of Q.

Now let us consider that a node equality condition is imposed on the instances of a path P. In this situation we have $Last(I_{P_1}) =_N Last(I_{P_2})$. Clearly, in this case, P has only one instance.   □

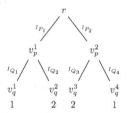

Fig. 7. An XML tree with two instances value equal for the path P and four instances for the path Q with I_{P_1} prefix of I_{Q_1}, I_{Q_2} and I_{P_2} prefix of I_{Q_3}, I_{Q_4}

Based on Example 9, we introduce the definition of our special tree, having *at most* two instances for each path in $\mathbb{P}$.

Definition 11 (Two-instance Tree)
Let $\mathcal{F}$ be a set of XFD. Let $\mathcal{T} = (t, type, value)$ be an XML document where the tree t, built according to the construction properties below, is called two-instance tree. Let $\mathbb{P}$ be the set of paths associated to $\mathcal{T}$, let $X \subseteq \mathbb{P}$, and let $E = (E_1, \ldots, E_n)$ be the equality list associated to X. We denote by $|Instances(P, t)|$ the number of instances of a path P in t.

CONSTRUCTION PROPERTIES:

1. For each $P \in \mathbb{P}$, $|Instances(P,t)|$ is at most 2 (I_1 or I_2) and when
   $|Instances(P,t)| = 2$:
   (a) we have $C/P[V] \in (C, X[E])^+$ iff $Last(I_1) =_V Last(I_2)$;
   (b) we have $C/P[E'] \notin (C, X[E])^+$ iff $Last(I_1) \neq_{E'} Last(I_2)$
2. For each $P \in \mathbb{P}$, $|Instances(P,t)| = 2$ except when:
   (a) $Last(P)$ is the root of t, or
   (b) by considering value equality, $|Instances(P,t)| = 2$ provokes the viola-
       tion of condition 1 for another path $Q \in \mathbb{P}$ with $P \prec Q$, or
   (c) $X[E] \rightarrow P[N]$ or
   (d) $Last(P) \in \Sigma_{att}$ and $Parent(P)$ verifies condition 2a, or 2b or 2c.   □

Lemma 6. Let $\mathcal{F}$ be a set of XFD. Let $\mathcal{T} = (t, type, value)$ be an XML document
where t is a two-instance tree. Let $\mathbb{P}$ be the set of paths associated to $\mathcal{T}$ and let
$X \subseteq \mathbb{P}$. The following properties hold for t:

1. If $C/P \in \mathbb{P}$ and $|Instances(C/P,t)| = 1$ then $C/P[E'] \in (C, X[E])^+$.

2. If $P, Q \in \mathbb{P}$, $P \preceq Q$ and there is an instance $I_P \in Instances(P,t)$, and
   instances I_{Q_1} and $I_{Q_2} \in Instances(Q,t)$ such that $I_P \preceq I_{Q_1}$ and $I_P \preceq I_{Q_2}$
   then $|Instances(P,t)| = 1$.

3. If $C/P \in \mathbb{P}$ then $C/P[E'] \in (C, X[E])^+$ iff $\mathcal{T} \models (C, (X[E] \rightarrow P[E']))$.   □

We now prove that the axiom system introduced in Definition 8 is complete.
In other words, given a set of XFD $\mathcal{F}$, by using our inference rules, we can derive
all XFD f such that $\mathcal{F} \models f$.

Theorem 3. If $\mathcal{F} \models f$ then $\mathcal{F} \vdash f$.   □

Proof: The proof is by contrapositive: we show that if $\mathcal{F} \nvdash f$ then $\mathcal{F} \nvDash f$.
Let $f = (C, (\{P_1[E_1], \ldots, P_n[E_n]\} \rightarrow \{P_{n+1}[E_{n+1}] \ldots P_{n+m}[E_{n+m}]\}))$. Then,
we consider that $X = \{C/P_1, \ldots, C/P_n\}$, $Y = \{C/P_{n+1}, \ldots, C/P_{n+m}\}$ and that
both X and Y are in a given $\mathbb{P}$. Let $E = (E_1, \ldots, E_n)$.
If $\mathcal{F} \nvDash f$ then there must be an XML document that satisfies $\mathcal{F}$ but does not
satisfy f. The proof consists in showing the existence of such a document.

Let us suppose an XML document $\mathcal{T} = (t, type, value)$ where t is a two-instance
tree defined on the set of paths $X = \{C/P_1, \ldots, C/P_n\}$.

Fact 1: $\mathcal{T} \models \mathcal{F}$
The proof is by contradiction. We suppose that $\mathcal{T} \nvDash g$, where g is an XFD
$(C, (\{Q_1[E_1'], \ldots, Q_k[E_k']\} \rightarrow Q_{k+1}[E_{k+1}']))$ in $\mathcal{F}$. From Definition 5, as $\mathcal{T} \nvDash g$,
we can deduce that there exist two projections $\Pi_M^1(\mathcal{T})$ and $\Pi_M^2(\mathcal{T})$ for the
branching path $M = \{C/Q_1, \ldots, C/Q_{k+1}\}$ in $\mathcal{T}$ such that:

$$\tau^1[C/Q_1, \ldots, C/Q_k] =_{E_i', i \in [1 \ldots k]} \tau^2[C/Q_1, \ldots, C/Q_k] \quad \text{and} \tag{6}$$

$$\tau^1[C/Q_{k+1}] \neq_{E_{k+1}'} \tau^2[C/Q_{k+1}]. \tag{7}$$

From (7) we have that $\Pi_M^1(\mathcal{T})[C/Q_{k+1}] \neq \Pi_M^2(\mathcal{T})[C/Q_{k+1}]$ and $|Instances$ $(Q_{k+1}, t)| = 2$. From Definition 11(1), we obtain:

$$C/Q_{k+1}[E'_{k+1}] \notin (C, X[E])^+. \tag{8}$$

From Definition 2, we know that the instances of two paths belonging to the same branching path match on their longest common prefix path. Formally, for all combination of paths Q_i and Q_j such that $1 \leq i \leq k+1$ and $1 \leq j \leq k+1$, we have:

Considering $\Pi_M^1(\mathcal{T})$:

$$isInst\_lcp(C/Q_i, \Pi_M^1(\mathcal{T})[C/Q_i], C/Q_j, \Pi_M^1(\mathcal{T})[C/Q_j]) = true \tag{9}$$

Considering $\Pi_M^2(\mathcal{T})$:

$$isInst\_lcp(C/Q_i, \Pi_M^2(\mathcal{T})[C/Q_i], C/Q_j, \Pi_M^2(\mathcal{T})[C/Q_j]) = true \tag{10}$$

and we can also determine the following special nodes for $1 \leq i \leq k$:

$$v_{i,k+1}^1 = Last(\Pi_M^1(\mathcal{T})[C/Q_i] \cap \Pi_M^1(\mathcal{T})[C/Q_{k+1}]) \quad \text{and} \tag{11}$$
$$v_{i,k+1}^2 = Last(\Pi_M^2(\mathcal{T})[C/Q_i] \cap \Pi_M^2(\mathcal{T})[C/Q_{k+1}])$$

From (9) and (10), together with the definition of $isInst\_lcp$ we know that positions $v_{i,k+1}^1$ and $v_{i,k+1}^2$ exist in t. We have to consider two cases:

(a) $v_{i,k+1}^1 = v_{i,k+1}^2$ (illustrated in Figure 8(a))

(b) $v_{i,k+1}^1 \neq v_{i,k+1}^2$ (illustrated in Figure 8(b))

We can easily show that it is always possible to choose for each $i \in [1 \ldots k]$, a path $C/R_i \in \mathbb{P}$ respecting the following property:

$$C/R_i[E'_i] \in (C, X[E])^+ \text{ and } Q_i \cap Q_{k+1} \preceq R_i \preceq Q_i \tag{12}$$

Now from property (12), the XFD $g = (C, (\{Q_1 [E'_1], \ldots, Q_k [E'_k]\} \to Q_{k+1} [E'_{k+1}]))$ in $\mathcal{F}$, and the axiom Branch Prefixing (Definition 8, axiom $A4$) we deduce the XFD $g' = (C, (\{R_1 [E'_1], \ldots, R_k [E'_k]\} \to Q_{k+1} [E'_{k+1}]))$. Next, we assume that if $C/\{R_1, \ldots, R_k\}[E'] \subseteq (C, X[E])^+$ then $C/Q_{k+1}[E'_{k+1}] \in (C, X[E])^+$. Indeed, by Definition 10 we know that $X[E] \to \{R_1 [E'_1], \ldots, R_k [E'_k]\}$. From this rule and g', we derive $X[E] \to Q_{k+1}[E'_{k+1}]$ by using the axiom Transitivity ($A3$). Thus, from Definition 10, we obtain $C/Q_{k+1}[E'_{k+1}] \in (C, X[E])^+$ which contradicts (8): $Q_{k+1}[E'_{k+1}] \notin (C, X[E])^+$. Thus, we conclude that $\mathcal{T} \models g$ for any $g \in \mathcal{F}$. In other words, $\mathcal{T} \models \mathcal{F}$.

Fact 2: $\mathcal{T} \not\models f$

Recall, from the beginning of our proof, that f is the XFD $(C, (X[E] \to Y[E'']))$. As $X[E] \subseteq (C, X[E])^+, \forall P_i \in X$ we have $Last(\Pi_M^1(\mathcal{T})[P_i]) =_{E_i} Last(\Pi_M^2(\mathcal{T})[P_i])$ for two given projections of M on $\mathcal{T}$. From our hypothesis, $\mathcal{F} \not\models f$ and so $C/Y[E''] \not\subseteq$

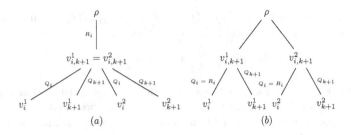

Fig. 8. Illustration of the two cases (a)$v^1_{i,k+1} = v^2_{i,k+1}$ and (b)$v^1_{i,k+1} \neq v^2_{i,k+1}$

$(C, X[E])^+$. Thus, there is at least one path $P \in Y$ having instances I_1 and I_2 such that $Last(I_1) \neq_E Last(I_2)$. We deduce that $\mathcal{T} \not\models f$.

In conclusion we have built a tree $\mathcal{T}$ such that $\mathcal{T} \models \mathcal{F}$ and $\mathcal{T} \not\models f$ which establishes the proof of Theorem 3.   □

Acknowledgments. We would like to thank Françoise Gire, Hicham Idabal, Béatrice Bouchou and Martin Musicante for initial discussions that gave rise to this work.

References

1. Abiteboul, S., Hull, R., Vianu, V.: Foundations of Databases. Addison-Wesley Publishing Company (1995)
2. Amavi, J., Halfeld Ferrari, M.: An axiom system for XML and an algorithm for filtering XFD. Tech. Rep. RR-2012-03, LIFO/Université d'Orléans (2012). http://www.univ-orleans.fr/lifo/rapports.php?lang=fr&annee=2012
3. Arenas, M., Libkin, L.: A normal form for XML documents. ACM Transactions on Database Systems (TODS) 29 No.1 (2004)
4. Bouchou, B., Cheriat, A., Halfeld Ferrari, M., Laurent, D., Lima, M.A., Musicante, M.: Efficient constraint validation for updated XML databases. Informatica **31**(3), 285–310 (2007)
5. Bouchou, B., Halfeld Ferrari, M., Lima, M.: Contraintes d'intégrité pour XML. visite guidée par une syntaxe homogène. Technique et Science Informatiques **28**(3), 331–364 (2009)
6. Bouchou, B., Halfeld-Ferrari, M., Lima, M.A.V.: A Grammarware for the Incremental Validation of Integrity Constraints on XML Documents under Multiple Updates. In: Hameurlain, A., Küng, J., Wagner, R., Liddle, S.W., Schewe, K.-D., Zhou, X. (eds.) Transactions on Large-Scale Data- and Knowledge-Centered Systems VI. LNCS, vol. 7600, pp. 167–197. Springer, Heidelberg (2012)
7. Castanier, E., Coletta, R., Valduriez, P., Frisch, C.: Public data integration with websmatch. In: BDA 2012 (2012)
8. Cecchin, F., de Aguiar Ciferri, C.D., Hara, C.S.: XML data fusion. In: DaWak. pp. 297–308 (2010)
9. Chabin, J., Halfeld Ferrari, M., Musicante, M.A., Réty, P.: Conservative Type Extensions for XML Data. In: Hameurlain, A., Küng, J., Wagner, R. (eds.) TLDKS IX. LNCS, vol. 7980, pp. 65–94. Springer, Heidelberg (2013)

10. Chirkova, R., Libkin, L., Reutter, J.L.: Tractable XML data exchange via relations. In: Proceedings of the 20th ACM Conference on Information and Knowledge Management, CIKM 2011. pp. 1629–1638 (2011)
11. Crane, G.: What do you do with a million books? D-Lib Magazine 12(3) (March 2006)
12. Gire, F., Idabal, H.: Regular tree patterns: a uniform formalism for update queries and functional dependencies in XML. In: EDBT/ICDT Workshops (2010)
13. Hartmann, S., Trinh, T.: Axiomatising Functional Dependencies for XML with Frequencies. In: Dix, J., Hegner, S.J. (eds.) FoIKS 2006. LNCS, vol. 3861, pp. 159–178. Springer, Heidelberg (2006)
14. He, Q., Ling, T.W.: Extending and inferring functional dependencies in schema transformation. In: Proceedings of the 2004 ACM CIKM International Conference on Information and Knowledge Management. pp. 12–21 (2004)
15. Kot, Ł., White, W.: Characterization of the Interaction of XML Functional Dependencies with DTDs. In: Schwentick, T., Suciu, D. (eds.) ICDT 2007. LNCS, vol. 4353, pp. 119–133. Springer, Heidelberg (2006)
16. Li Lee, M., Ling, T.-W., Low, W.L.: Designing Functional Dependencies for XML. In: Jensen, C.S., Jeffery, K., Pokorný, J., Šaltenis, S., Bertino, E., Böhm, K., Jarke, M. (eds.) EDBT 2002. LNCS, vol. 2287, p. 124. Springer, Heidelberg (2002)
17. Liu, J., Vincent, M.W., Liu, C.: Functional dependencies, from relational to XML. In: Ershov Memorial Conference. pp. 531–538 (2003)
18. Pankowski, T.: Reconciling Inconsistent Data in Probabilistic XML Data Integration. In: Gray, A., Jeffery, K., Shao, J. (eds.) BNCOD 2008. LNCS, vol. 5071, pp. 75–86. Springer, Heidelberg (2008)
19. Shahriar, M.S., Liu, J.: Preserving Functional Dependency in XML Data Transformation. In: Atzeni, P., Caplinskas, A., Jaakkola, H. (eds.) ADBIS 2008. LNCS, vol. 5207, pp. 262–278. Springer, Heidelberg (2008)
20. Vincent, M.W., Liu, J., Liu, C.: Strong functional dependencies and their application to normal forms in XML. ACM Trans. Database Syst. 29(3), 445–462 (2004)
21. Vincent, M., Liu, J., Mohania, M.: The implication problem for 'closest node' functional dependencies in complete XML documents. Journal of Computer and System Sciences 78(4), 1045–1098 (2012)
22. Wang, J., Topor, R.: Removing XML data redundancies using functional and equality-generating dependencies. In: ADC 2005: Proceedings of the 16th Australasian database conference. pp. 65–74. Australian Computer Society Inc., Darlinghurst, Australia (2005)
23. Wu, X., Ling, T.W., Lee, M.L., Dobbie, G.: Designing semistructured databases using ORA-SS model. In: Proceedings of the 2nd International Conference on Web Information Systems Engineering, WISE (1) (2001)
24. Zhao, X., Xin, J., Zhang, E.: XML functional dependency and schema normalization. In: HIS 2009: Proceedings of the 9th International Conference on Hybrid Intelligent Systems. pp. 307–312 (2009)

Author Index

Printed in the United States
By Bookmasters